Water Allocation in Rivers under Pressure

Water Trading, Transaction Costs and Transboundary Governance in the Western US and Australia

Dustin Evan Garrick

McMaster University, Canada

Cheltenham, UK • Northampton, MA, USA

Published by
Edward Elgar Publishing Limited
The Lypiatts
15 Lansdown Road
Cheltenham
Glos GL50 2JA
UK

Edward Elgar Publishing, Inc.
William Pratt House
9 Dewey Court
Northampton
Massachusetts 01060
USA

Paperback edition 2017

A catalogue record for this book
is available from the British Library

Library of Congress Control Number: 2014959462

This book is available electronically in the **Elgar**online
Social and Political Science subject collection
DOI 10.4337/9781781003862

ISBN 978 1 78100 385 5 (cased)
ISBN 978 1 78100 386 2 (eBook)
ISBN 978 1 78643 599 6 (paperback)

Typeset by Servis Filmsetting Ltd, Stockport, Cheshire
Printed and bound in Great Britain by TJ International Ltd, Padstow

Contents

Table of conversions

	Litres	Cubic metres	Megalitres	Acre feet
1 litre	1	1×10^{-3}	1×10^{-6}	8.1×10^{-7}
1 cubic metre	1000	1	1×10^{-3}	8.1×10^{-4}
1 megalitre*	1 000 000	1000	1	8.1×10^{-1}
1 acre foot	1 233 482	1233	1.233	1

*1 Gigalitre is 1000 megalitres

Cubic metres are the international standard. *Acre feet* tend to be used in the US. *Megalitres* and *gigalitres* are used in Australia.

Preface and acknowledgements

The origin of this book can be traced to the summer of 2002. I spent a year in the Sonoran Desert of Arizona for my Master's degree at Columbia University, examining efforts to manage groundwater overdraft in an arid and unforgiving landscape. Also, 2002 was one of the driest years in the Colorado River since at least 1977, casting Arizona's groundwater management challenges into sharp relief against the backdrop of climate change, competition and conflicts over water across the Intermountain West of the US. It is now 2015, and the Colorado River remains gripped by drought and under increasing pressure as Lake Mead – one of its primary reservoirs – has reached historic lows. In the ensuing period, I have travelled the intellectual terrain of property rights theory, institutional economics and comparative public policy to understand the evolution and performance of water allocation policy in stressed rivers of the Western US and Australia.

This book compares water allocation reform in three stressed rivers – the Colorado and Columbia in North America, and the Murray–Darling in Australia – that have attempted similar paths toward sustainable and adaptive water management by experimenting with water trading and river basin governance reforms. Investigating and comparing these institutional pathways has been the adventure of many lifetimes. My journey has spanned from the terminus of the 336-mile Central Arizona Project canal to the fields of the Mexicali Valley in the Colorado River; from the dry tributaries of Flathead Lake, Montana to the restored salmon habitat of the John Day River, Oregon, a tributary of the Columbia River; and from the Murray Mouth to flashy tributaries of the Darling River. In the process, this work has attracted interest and opened scholarly and policy networks. It has benefited immeasurably from a number of workshops, collaborations and presentations in Australia, China, the UK (Cambridge and Oxford), the US (Harvard, California–Berkeley, Indiana), South Africa, Germany (Humboldt University) and Sweden, to name just a few of the venues where the work in this book has been presented and debated.

I have accrued many debts to those who have offered guidance, expertise and encouragement during a journey that also proved arduous and fraught with challenges. Thank you to Tanya Heikkila and the Biosphere 2 team for framing the water sustainability challenges and offering the encouragement and theoretical lenses to continue this work. John Horning taught me to think like a beaver and appreciate rivers as the lifeblood of the Western US ecology, culture and economy; his and Terry's La Jencia ranch offered the retreat where many of this book's ideas came into focus. Kathy Jacobs provided the opportunity to connect science and policy in the Colorado River during an unprecedented transition in the Basin with added thanks to Peter Culp and those at the Bureau of Reclamation (led by Terry Fulp) who sought to integrate climate science into water management despite significant professional risks attached with doing so.

My PhD research in the Columbia River Basin would not have been possible without the steady guidance and committed mentorship from Carl Bauer and Edella Schlager, the co-chairs of my dissertation committee. I could not have asked for a better mix of intellectual perspectives and professional advice. Several other colleagues at the University of Arizona contributed to a vibrant intellectual environment for this work, including Bonnie Colby, Chris Scott, Paul Robbins, Robert Glennon, David Tecklin and Sharon Megdal. Bruce Aylward provided a unique blend of academic advice and practical experience. He guided my path in the Columbia to examine innovations in market-based environmental water acquisition and collaborative governance that were the envy of the rest of the Western US and gaining the attention of the rest of the world. He also showed me that the world was bigger than the Western US, organizing a meeting in Brisbane to exchange lessons between the Pacific Northwest and the Murray–Darling Basin, opening a path that I would pursue after finishing my PhD. The practitioners (more than 50) in the Columbia River Basin led by Andrew Purkey of the Columbia Basin Water Transactions Program were generous with their expertise and friendship. The list is too long to acknowledge, but many had a genuine interest in my work and helped me in my attempts to bridge the divide between research and practice (a partial list includes Molly Whitney, Kacy Markowitz, David Pilz, Morgan Case, Amanda Cronin, Bob Barwin, Scott McCaulou, Rankin Holmes, Brett Goldin, Tod Heisler, Lisa Pelly, Steve Parrett, Brianna Randall, Kim Schonek, Kevin Scribner, Laura Ziemer and Stan Bradshaw).

I spent a year in Australia (2010–11) on a Fulbright scholarship. This provided a chance to compare institutional reforms in the Western US and the Murray–Darling Basin – the latter being a grand experiment in water reform to 'claw' water back for the environment. Thank you to sponsors at the Commonwealth Scientific and Industrial Research Organisation (CSIRO) Ecosystem Sciences (Jeff Connor) and Australian National University (Daniel Connell, Jamie Pittock and Quentin Grafton) as well as my other hosts and colleagues throughout Australia, especially Rosalind Bark (now at Leeds), Onil Banerjee (now at the Inter-American

Development Bank), Henning and Vibeke Bjornlund, Adam Loch and Sarah Wheeler at the University of South Australia; Cathy Robinson, Anthea Coggan and Stuart Whitten of CSIRO; Mike Young and Adam Webster of the University of Adelaide; Graham Marshall of the University of New England; Erin O'Donnell and Avril Horne of University of Melbourne; Lin Crase of Latrobe University; multiple colleagues at Australian state agencies and the Murray–Darling Basin Authority (including Tony McLeod and Jason Alexandra); and Mark Siebentritt. In 2013 I returned as a visiting fellow to the Australian National University to complete Chapter 5 of this book.

Additional thanks go to my mentors and colleagues at the University of Oxford where I completed a research fellowship from 2011 and 2013. Special thanks to David Grey, Jim Hall, Simon Dadson, Rob Hope, Mark New and Karis McLaughlin. David Grey instilled an ethic of putting science into practice to help solve the pressing water security challenges of our time; he reminded me that the rural settlements of the Western US and Australia were once inhospitable and impoverished regions – a 'difficult hydrology' that many of the world's poor still face. Since 2014, McMaster University has provided an ideal home to advance this project through discussions with Mark Sproule-Jones and other colleagues in the Department of Political Science, Booth School of Engineering Practice and McMaster Water Network.

A number of academic colleagues have generously reviewed portions of this book, with special thanks to Laura McCann (Missouri) who offered critical feedback during my PhD and beyond. I received scholarly feedback from many of the colleagues noted above, with a special acknowledgement of the chapter reviews by Bruce Aylward, Rosalind Bark, Carl Bauer, Daniel Connell, Graham Marshall, Laura McCann, Amy McCoy, Edella Schlager and David Zetland. Karen Devivo and Celeste Robitaille completed the thankless copy-editing and formatting tasks long after my eyes had been blinded by revisions. Jimmy Mack and Kira Moor helped with maps at various stages.

I have reserved the last notes of gratitude for those whose support and friendship have buoyed me during the most challenging moments and inspired me to be a better friend, brother, son, father, and husband. Thank you to Amy McCoy and Mike Sawicki for sharing this journey and deepening our friendship in the process; Nadine, Steve, Emily and Al for the brotherly and sisterly support, and my nieces, Amaya and Mikayla. Special thanks to my mom, Hannah, and dad, Arthur, for their wisdom, love and steadfast encouragement, and to Mark and Phyllis for treating me as their own. At the end of this journey, I do not have words to thank and honor Heather, for being by my side in every way, and Skylar for being the joy of our lives and the beacon of the future.

Abbreviations

AF	Acre Foot
BPA	Bonneville Power Administration
CBWTP	Columbia Basin Water Transactions Program
CEWO	Commonwealth Environmental Water Office
CFS	cubic feet per second
COAG	Council of Australian Governments
COW	Coase, Ostrom, Williamson
CPRs	common pool resources
CSIRO	Commonwealth Scientific and Industrial Research Organisation
C-WON	Coase, Williamson, Ostrom, North
DRC	Deschutes River Conservancy
ESA	Endangered Species Act
FCRPS	Federal Columbia River Power System
GAO	Government Accountability Office
GDP	gross domestic product
GL	gigalitre
IAD	institutional analysis and development framework
IAG	Independent Audit Group (Murray–Darling Basin)
ICA	institutional collective action
IWRM	integrated water resources management
LDS	lower division states (Colorado River)
MAF	million acre feet
MDB	Murray–Darling Basin
MDBA	Murray–Darling Basin Authority
MDBC	Murray–Darling Basin Commission
MDBMC	Murray–Darling Basin Ministerial Council
ML	megalitre
NFWF	National Fish and Wildlife Foundation
NGO	non-governmental organization
NPCC	Northwest Power and Conservation Council (Columbia River)
NSW	New South Wales
NWC	National Water Commission (Australia)

NWI	National Water Initiative (Australia)
O&M	operations and maintenance
OECD	Organisation for Economic Co-operation and Development
QLE	qualified local entity (Columbia River)
SCM	Southern Connected Murray
SDL	sustainable diversion limit
TLM	The Living Murray
UDS	upper division states (Colorado River)

1. Water allocation in rivers under pressure: a large-scale collective action dilemma

> Water rights transfers would increase the benefits gained from the use of water and would tend to delay or make unnecessary the construction of new sources of supply . . . [but] the fact that no two water rights are identical ... will prevent the development of a market in water rights comparable to the auction market of a stock or commodity exchange. (National Water Commission, 1973: 260, identifying the benefits and constraints on water trading)

> The current system is clunky: it's often difficult to get approvals, and protections are not always effective. We will need to develop a more streamlined approach, while taking into account that the water market is much more nuanced than, say, a market for plywood. How and when water moves through the system matters, so rules are needed to facilitate trading and to ensure that it doesn't harm other users or the environment. (Ellen Hanak, *New York Times*, 29 June 2014, on the state of water rights trading 40 years later in the context of California's drought, 2011 to 2014, and ongoing as of 2015)

INTRODUCTION

Sustainable and secure access to freshwater is a defining challenge for society. This ancient challenge is becoming increasingly difficult because many of the world's rivers and aquifers are under pressure. Almost 4 billion people will live in river basins experiencing severe water stress by 2050 according to the baseline scenario of the most recent Organisation for Economic Co-operation and Development (OECD) environmental outlook study (OECD, 2012). This is not a distant challenge, however. A 2010 study in *Nature* concluded that 80 per cent of the global population faces threats to water security, particularly in regions with 'intensive agriculture and dense settlement' and where 'water scarcity accentuates threats to drylands' (Vorosmarty et al., 2010: 556). In this context, the World Economic Forum has identified water crises among the top societal risks in its annual survey of global leaders from 2011 to 2015. Recognition of these threats is growing, and water crises topped the 2015 global survey as the highest impact risk (World Economic Forum, 2015). Water crises are also positioned as a 'center of gravity' that binds together agriculture, energy, and ecosystems and connects local and global processes through trade, demographic patterns and climate change (World Economic Forum, 2013).

Intensified competition for scarce and variable water supplies requires water allocation institutions – shared rules and norms to establish and allocate property rights to water. These institutions determine how to allocate water across competing uses, reduce and share risks, manage conflicts, and adapt in the face of stress and change. Water allocation institutions include a bundle of property rights governing who can access, withdraw and consume water, and under which conditions. Water allocation institutions also define who manages and reallocates water, and how water's multiple public goods will be provided and sustained (Schlager and Ostrom, 1992; Schlager and Blomquist, 2008; Meinzen-Dick, 2014).

It has been 40 years since the US National Water Commission identified voluntary water transfers as part of the 'future of water policy' (NWC, 1973). Yet the transition to more sustainable and adaptive water allocation institutions remains elusive. This book investigates why progress has proven slower and more uneven than anticipated, and it identifies institutional choices that enhance or reduce the capacity to adapt and accelerate reforms. As the number of uses and values (cultural, economic and ecological) proliferates, so do the stakeholders involved in water allocation and governance, spanning diverse water users, different sectors (irrigation, cities, energy), national and subnational governments, river basin organizations and international organizations, among others. The precise mix of stakeholders is an empirical question, but water allocation institutions increasingly involve decisions with impacts and implications that cut across multiple political boundaries. This heightens the calls for institutions that not only enhance efficiency, equity, sustainability and adaptive capacity, but also promote integration and coherence across multiple values, scales and jurisdictions.

Principles for 'good water governance' and policy prescriptions to achieve more sustainable and adaptive water allocation systems have been developed in tandem with the growing perception that there is a (global) water crisis (Figure 1.1). The emerging paradigms for water governance and water allocation reform have been anchored in local experiences, which have produced a diverse landscape of allocation rules with different mixtures of three ideal types: community-based management (users), markets (water rights trading) and governments (central governments) (Meinzen-Dick, 2007). Despite the diversity of rules and norms and the importance of local context, global experts and epistemic communities have converged

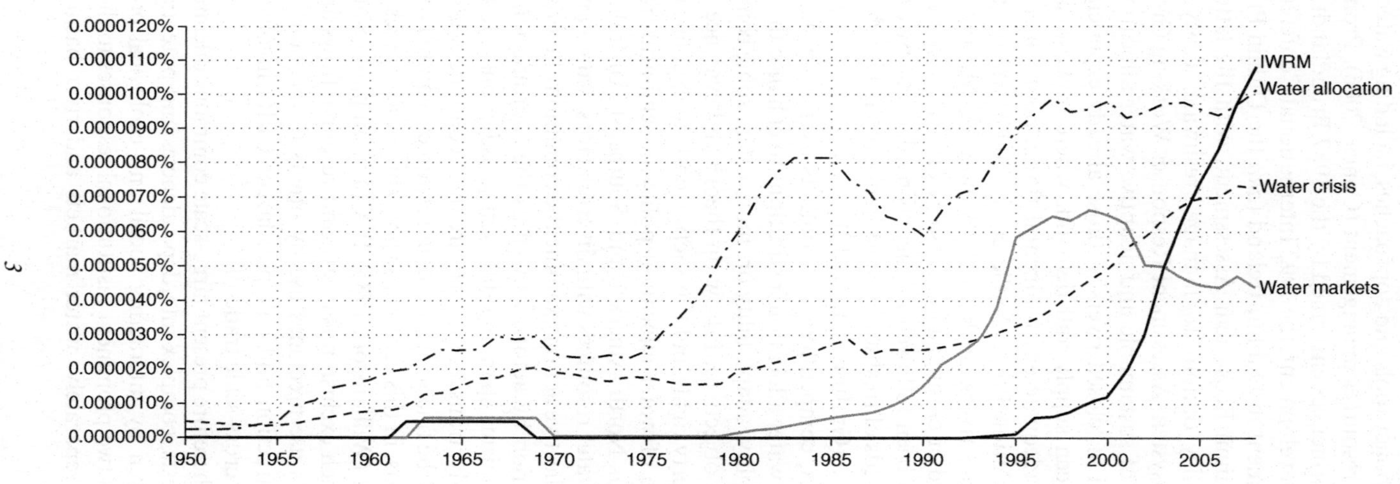

Figure 1.1 Google 'N-gram' trend analysis of phrase appearance in books 1950–2008

on some specific policy tools and approaches, including water markets and integrated water resources management (Conca, 2006). Prominent examples include the principles enshrined by the 1987 Brundlandt Commission on Sustainable Development, the 1992 International Conference on Water and the Environment in Dublin, Ireland (*aka* the Dublin Principles) and the 2005 Millennium Ecosystem Assessment, with the latter dedicating a chapter to policy options to manage freshwater ecosystem services (Millennium Ecosystem Assessment Responses Working Group, 2005).

The challenge of sustainable and adaptive water allocation is particularly pronounced in 'closed rivers' – those already experiencing chronic imbalances between supply and demand, where downstream needs are unmet due to inadequate environmental flows, poor water quality and/or insufficient dilution flows (Molle et al., 2007). Many of the world's iconic mid-latitude rivers, such as the Colorado, Murray–Darling and Yellow, are closed or closing; a long history of hydropower and irrigation development has been followed by increasing demand by cities, industry and the unmet needs of freshwater ecosystems, often with sharply reduced outflows to the sea (Grafton et al., 2012).

In closed rivers, competition and conflicts over water have prompted efforts to reform water allocation institutions to enhance their equity, efficiency, sustainability, adaptability or, more often, some complex mixture of these multiple objectives. I draw on theory and evidence about collective action, property rights and transaction costs to examine the evolution and performance of water allocation policy in semi-arid transboundary rivers of Western North America and Southeast Australia: large-scale 'common pool' water resources with diverse actors, values and interests.

Theories of collective action examine the conditions favoring or hindering self-organization by actors, as well as the coordination and conflict resolution mechanisms spanning different jurisdictions. However, collective action is costly; transaction costs are the 'economic equivalent of friction' and refer to the financial and other resources required to develop and change institutions, in this case, those governing water access, withdrawal and management (Williamson, 1985: 2). Common pool resources (CPRs) are those for which exclusion is costly and resource flows are 'subtractable': one person's use precludes use by others in the short term, posing a risk of overextraction and collapse; larger CPRs involve more users, jurisdictions and cross-scale trade-offs.

Closed rivers therefore present large-scale commons dilemmas, comprising diverse interests, complex interdependencies and potential for threshold changes driven by interacting social and ecological pressures. This book investigates two prominent institutional responses to these dilemmas often examined separately: water markets and river basin governance.

I use evidence from in-depth interviews and questionnaires, historical archives and quantitative and qualitative measures of transaction costs to compare institutional design and performance in three river basins: the Colorado, Columbia and Murray–Darling. All three are on the leading edge of global water challenges yet have experienced varying levels of success and diverging trajectories of reform.

THE WATER CRISIS: A CRISIS OF FRAGMENTED GOVERNANCE – LESSONS FROM THE COMMONS

The 'global water crisis' has been attributed to a variety of factors. Global population has doubled since the 1960s, and hit 7 billion in 2012, with at least 2 billion more people projected by 2050.[1] This population is increasingly urban and adopting resource-intensive diets that could require a substantial increase in food production by 2050. The environment has been recognized as a legitimate and economically important water user, yet aquatic habitat associated with 65 per cent of river discharge is compromised (Vorosmarty et al., 2010). Climate change is a threat multiplier with water-related impacts superimposed on chronic pressures and resource scarcity. A 2°C increase in global average temperature is projected to increase the global population living in conditions of absolute water scarcity (Schewe et al., 2014).

In this context, the OECD surveyed water governance trends in its member countries and concluded that the water crisis is fundamentally a crisis of governance and policy fragmentation, requiring collective action by water users, sectors, jurisdictions and stakeholders at all levels (OECD, 2011). This is not the first time the water crisis has been equated with a crisis of fragmented governance, as it followed previous declarations by the Global Water Partnership and UN-Water. Nevertheless, the scale and complexity of the challenge are increasing in the context of growing demands, climate variability and resource interdependencies across interrelated food, energy and water systems. The water crisis is therefore a prime example of a large-scale common pool resource governance dilemma.

The governance challenges for water have been framed in terms of the 'tragedy of the commons' (Hardin, 1968). As noted, common pool resources (CPRs) have unique characteristics: high costs of exclusion, making it difficult to prevent new users from accessing the resource; and subtractability, meaning that consumption of resource flows by one user prevents others from doing so until the resource is replenished. Overuse may lead to collapse when the resource stock components of the CPR are impaired and the system loses its capacity to replenish resource flows, for example when watershed functions are compromised or when groundwater extraction leads to subsidence.

Water governance challenges arise in part from water's status as a complex 'economic good' with interdependent private and public values and uses each involving different forms of property rights and coordination institutions. Michael Hanemann (2006) considered the thesis that water is different from traditional economic commodities. Water and water infrastructure have public good characteristics; many benefits, such as environmental flows and flood control, are indivisible for most practical purposes and therefore potentially underprovided because people benefit regardless of their contributions. Water is also distinct due to its complex supply characteristics: variability in space and time, mobility, the capital intensity of large water storage and distribution, and heterogeneity across different sources (groundwater, surface water, and so on) and uses. Together these supply and demand characteristics require a level of collective action – and hence transaction costs – that 'simply does not arise with most other commodities' (Hanneman, 2006: 87). As a consequence, property rights to water are comparably costly to define and manage across interdependent private and public uses.

Despite the costs and challenges involved with governing the commons, lessons have emerged from both failed and successful attempts to do so. Nearing 50 years since Hardin's (1968) provocative thesis about the tragedy of the commons, the accumulated empirical evidence demonstrates that collapse is not inevitable despite the collective action dilemmas associated with a multidimensional resource such as water. A body of theory on collective action and institutional change has identified the potential to govern commons sustainably under certain conditions, such as shared norms, small groups, low inequality and clearly defined and understood boundaries around the resource (Cox et al., 2010: 38; Ostrom, 1990). Larger systems such as transboundary river basins pose added complexity, costs and challenges. In an agenda-setting paper in *Science* in 1999, late Nobel laureate Elinor Ostrom and colleagues (Ostrom et al., 1999) diagnosed the special features of large-scale[2] commons dilemmas:

1. The challenge of 'scaling up': the cost of collective action often rises with the number of actors.
2. Cultural diversity: diversity involves multiple values and, potentially, less scope for shared interests.
3. Different CPRs are linked: large CPRs involve complex connections between different systems (for example, climate, water, forests).

4. Accelerating rates of change: global change and disturbances present robustness challenges.
5. Unanimity as the default decision-rule: inertia and lowest common denominator approaches are prevalent when unanimous decisions are needed.
6. One blue planet: unlike local community-based systems, limited alternatives are available if we fail to steward the planet or major systems within it.

INSTITUTIONAL DIVERSITY AND A TALE OF TWO PANACEAS: WATER MARKETS AND RIVER BASIN GOVERNANCE

Institutions are the 'prescriptions that humans use to organize all forms of repetitive and structured interactions' (Ostrom, 2009: 3). 'Institutional diversity' refers to the potential for these prescriptions to be devised and matched to local and regional circumstances. In her discussion of the rules used to allocate CPRs, for example, Ostrom (2009) notes 112 potential allocation rules based on combinations of *allocation formulas* (n = 8), such as a volume of water per year, and the *basis for allocation* (n = 14), such as historic use as the basis for appropriative water rights (ibid: p.222). For large-scale CPRs, multilevel institutions add layers of complexity by combining rules associated with user self-governance, markets, states and even regional and global governance arrangements, with varying levels of coherency and coordination, across types and scales. In this context, it is tempting to simplify the complexity. Institutional blueprints or 'panaceas' have been proposed as a one-size-fits-all (or most) cure for complex CPR governance dilemmas. At the same time, there is a growing call to take context seriously in devising and adapting broad principles to local circumstances, as it is hard to predict how the imposition or removal of a single rule or set of rules will affect the performance of the overall governance arrangements.

Ruth Meinzen-Dick (2007) wrote about three prominent panaceas applied in water institutions – states, markets, and users or community-based management[3] – and argues that they should be viewed as mutually reinforcing pillars of an 'institutional tripod' rather than mutually exclusive alternatives. To these three types, river basin organizations and integrated water resources management[4] have emerged as a 'nirvana concept' (Molle, 2008), and arguably a fourth panacea that overlays the others (Thiel, 2014). River basin governance addresses the widening range of interests connected to water; the boundaries of rivers or watersheds are upheld as the preferred management unit to internalize the externalities of water withdrawal and use.

In the water-stressed and closed river basins of the world, water markets and river basin governance are two panaceas to manage water scarcity and climate variability, often promoted, designed and implemented independently (with the Murray–Darling of Southeast Australia among the first to explicitly coordinate the two).

Panacea 1: Water Markets

Water markets have received attention since the 1970s as an institutional response to water stress and associated governance and market failures. Water markets are a form of cap-and-trade environmental and natural resource allocation policy. A cap on water diversions establishes limits, ideally based on sustainability criteria. The allocation of water entitlements creates an initial distribution of property rights based on political contests and principles of efficiency and fairness. Trading rules allow voluntary reallocation to enhance efficiency. The appeal of water markets lies in their potential to address competition and manage conflicts over water by using voluntary transactions between willing buyers and sellers to reallocate water. Equally important is the potential for water markets to delay or even avoid costly new supply infrastructure by facilitating more efficient use of existing resources.

From an institutional design standpoint, there are three basic elements underpinning water markets, and related market-based reallocation mechanisms: a cap, an initial allocation, and trading rules to govern reallocation. The establishment of a cap limits the cumulative supply, or consumptive pool, available for water use in a given region. Such a cap is a regulatory choice, triggered by perceptions of scarcity, and ideally based on sustainable limits. A cap means that new water uses cannot be met without a temporary or permanent reduction in existing uses. The cap therefore provides economic incentives to reallocate existing entitlements (because new permits are no longer a viable option to satisfy emerging demands). Increasingly, the river basin is used as the spatial unit for defining a nested set of caps in which local limits are developed within basin-wide constraints to reduce the social and environmental impacts of changing water use patterns. In the absence of regulation, physical limits on renewable water supplies are an implicit cap and can result in the dewatering of rivers and impairment of aquifers. Indeed, the physical closure of basins often triggers the establishment of environmental reserves. Water markets also depend on institutional reforms to establish private, tradable and enforced property rights to water, as well as regulatory safeguards to address social and environmental concerns. Therefore, the second element of water market design is water rights reform. Tradable water rights – the ability to buy and sell well-defined water use entitlements – enable voluntary reallocation among water users. In theory, price signals are used in lieu of administrative or court decisions to cue shifts in water use patterns that maximize productivity as availability and preferences change.

Water markets and cap-and-trade water allocation reforms are hardly novel or recent. It has been 40 years since the National Water Commission encouraged reallocation to 'increase the benefits gained from the use of water and . . . delay or make unnecessary the construction of new sources of supply' (NWC, 1973; Congressional Research Service, 2009). Even in 1973, market-oriented reallocation hinged on effective governance and robust institutions. Proposed water allocation reforms implied a single institutional transition, or 'set-up' period, to complete market-enabling reforms by creating water rights registries, adopting trading rules, reducing legal barriers, and establishing water supply projects to handle storage and distribution (Congressional Research Service, 2009). These early arguments for voluntary water reallocation via market mechanisms in the 1970s were followed in the early 1990s by 'free market environmentalism', in which private, tradable and exclusive property rights were the solution to resource scarcity and environmental pollution problems as part of a natural progression toward more economically efficient resource allocation institutions (Anderson and Leal, 2001); for a critical view of the logic of efficient property rights to water, see for example Rose (1990). The scholarly community has developed an abiding interest in water markets with major works assessing the emergence (Saliba and Bush, 1987; Howe et al., 1990; National Research Council, 1992; Rosegrant and Binswanger, 1994; Easter et al., 1998), maturation (Bennett, 2005; Olmstead, 2010; Grafton et al., 2011) and future (Anderson et al., 2012; Maestu, 2013; Glennon, 2010; Easter and Huang, 2014) of water markets. A profound gap, however, has appeared between the initial logic of environmental markets and the reality on the ground. In the intervening years, on-the-ground experience has revealed that cap-and-trade water allocation reforms are more complex than initially envisioned (Neuman, 2004).

Critics of free-market environmentalism bemoaned early prescriptions as 'over-simplistic, misleading and hyperbolic' due to the 'failure of markets to allocate effectively environmental resources because of information costs, externalities, public goods and strategic behaviour' (Blumm, 1992: 372). The unfulfilled promise of some experiments with water markets and the challenges of using voluntary water transactions to resolve conflicts between public and private uses led Bauer (2004) to equate market reforms with a siren song that uses deceptive allure to trap its victims. Writing about the use of market-like water rights acquisition for environmental restoration in the US Pacific Northwest, Neuman (2004) captures this disconnect between the promise and progress succinctly: 'that was then, and this is now'. A more nuanced and mature conception of market environmentalism has developed in this period, comprising five interrelated elements: privatization, commercialization, environmental valuation, marketization and liberalization of governance (Bakker, 2014).

A pragmatic middle ground between unqualified success and abject failure has emerged in which water markets and voluntary water transactions have an important, yet complementary, role to play in enhancing the flexibility and adaptability of water allocation institutions. On the one hand, it is evident that water markets play an important role in water allocation in societies, economies and environments as diverse as Australia, Canada, China, Chile, South Africa and the Western US. On the other, the trading activity remains a fraction of the potential anticipated in most of these regions as recently as the early 1990s. The Australia experience in the Southern Murray Darling is the exception – where up to 40 per cent of water allocations in the Southern Murray are traded in a given year (Grafton and Horne, 2014). However, water trading in the Murray–Darling casts the lack of activity in the other semi-arid regions into sharp relief. Despite intense competition for water across agriculture, cities and ecosystems, for example, the California water market has grown only to five per cent of total water use by 2012 due to a range of complex regulatory and infrastructure issues (Hanak and Stryjewski, 2012).

Realizing the benefits of water trading depends on strong institutions and ongoing institutional adaptation from the community to river basin level and beyond. This nuanced governance view of water markets is epitomized by the comments from Ellen Hanak promoting voluntary water transfers as an important response to shortages during California's severe drought conditions in 2014. It also echoes early observations in the Southern Murray–Darling. Writing after the first decades of market-oriented institutional reforms in the Murray–Darling, Challen (2000) concluded that water markets cannot be considered as 'self-maintaining' allocation options that operate smoothly after a single set-up period. Instead, water markets require continuous and sequentially more complex institutional transitions (Carey and Sunding, 2001; Garrick et al., 2013). The evaluative criterion of water allocation institutions in this context must be multidimensional and dynamic as captured by the concept of adaptive efficiency, as discussed in Chapter 2.

Panacea 2: River Basin Governance and IWRM

The pervasive externalities and interdependencies associated with water allocation require multi-level and transboundary governance arrangements for coordination, conflict resolution and/or collaboration across political borders. The river basin scale has been upheld as the natural unit for management. In the Murray–Darling, for example, water markets have been accompanied by several decades of river basin planning and interstate agreements. River basin governance contributes directly to market-enabling institutional frameworks through auditing, capping, allocating and trading water rights; it also addresses a range of public goods and other water governance objectives related to sustainability and adaptation. The Murray–Darling therefore demonstrates that water rights reforms are necessary but insufficient for sustainable and adaptable water allocation policy. Robust and adaptive governance arrangements are needed for multi-jurisdictional planning and conflict resolution that combine water markets with strong local, state, federal and river basin institutions.

The coordination of upstream–downstream trade-offs in river basins is a challenge as old as civilization, yet river basin closure heightens interdependencies, as outlined below. As a result, several influential international bodies have promoted river basin governance, and its close cousin, integrated water resources management (IWRM), which together form another prominent panacea, revealed by the n-gram in Figure 1.1. Close scrutiny shows a spike in attention to IWRM in the mid-1990s, which coincided with the agenda and principles set out at the Dublin Conference on Water and the Environment. The Dublin Principles of 1992 were arguably the high-water mark of attention and commitment to the river basin as the preferred scale for water management. The action agenda called for efforts to protect aquatic ecosystems and noted that 'integrated management of river basins provides the opportunity to safeguard aquatic ecosystems, and make their benefits available to society on a sustainable basis' (ICWE, 1992). Even more noteworthy was the assertion that the river basin is the 'most appropriate geographical entity' for planning, management and conflict resolution. Much like the early ideas about water markets, however, for IWRM and river basin governance, that was then and this is now.

Scholars and practitioners have noted a range of justifications for developing or retaining water management at other scales; political jurisdictions are socially constructed and contested in relation to diverse values and interests, with different politics and economies of scale (Thiel, 2014; Blomquist and Schlager, 2005). The critiques of river basin governance and IWRM fall under four main categories (Thiel, 2014). First, the physical boundaries of catchments, watersheds, aquifers and river basins are not always clearly specified, particularly for large basins. As one example, for most practical and management purposes, the Darling River is a distinct system from the Murray, and several major tributaries of the Darling only contribute to the Murray in wet years. The situation for aquifers is even more complex, with several layers of shallow and deep aquifers overlaying each other with varying levels of interactions in parts of the Colorado River. Second, even when the biophysical boundaries are clear, problems of institutional fit exist when political borders are not aligned with the basin and aquifers (Mostert et al., 2008; Moss, 2004). Third, these mismatches are not inherently bad because many water management issues are addressed at other scales due to the politics involved (Molle, 2009, 2008). This aligns with another popular prescription which gained popularity in the early 1990s in the context of the European Union: subsidiarity, or the notion that decisions should be taken at the lowest level possible, but no lower (Marshall, 2007). As a consequence of the first three reasons, efforts to achieve comprehensive and integrated management at the river basin scale are constrained by transaction costs. As Schlager and Blomquist (2008) note, institutional development is costly and human rationality is bounded.

As with water markets, a middle ground has emerged between the extremes of abject failure and unmitigated success in the form of polycentric governance arrangements. Polycentric governance arrangements feature multiple independent centers operating in conflict, competition and/or cooperation with one another (Ostrom et al., 1961; Andersson and Ostrom, 2008; McGinnis and Ostrom, 2012); the transaction costs of coordination are a limiting factor affecting the degree of coherence and integration, both within (horizontal) and across (vertical) levels of governance. Marshall (2005) elaborates the twin concepts of subsidiarity and complementarity to convey the assignment and coordination of governance tasks across levels. *Subsidiarity* refers to the principle of assigning governance tasks at the lowest level possible, while *complementary* higher-level institutions are needed to coordinate tasks that span multiple political boundaries due to externalities or economies of scale (Marshall 2005). Efforts to scale up water markets and sustainable water allocation involve a nested set of formal and informal institutions to provide multiple water-related public goods; however, the costs and politics of institutional development mean that that an overarching river basin authority may not be either necessary or desirable.

PROBLEM STATEMENT

Water allocation in closed rivers presents 'wicked problems' (Rittel and Webber, 1973). The framing of the problem is in dispute, solutions have unintended consequences, and positive feedback contributes to lock-in, which makes learning and subsequent adaptations costly. The two panaceas introduced above share a similar arc – from promise to pitfalls to pragmatism – which reveals a gap between the theory and evidence of institutional change in water allocation and underscores that institutions are not free. Transaction costs are high, often prohibitively so, contributing to path dependent and incremental change that may fail to keep pace with society's evolving preferences and values. Transaction costs[5] refer broadly to the resources (money, staff, time and so on) required for collective action to develop, implement and adjust property rights and adopt institutional changes. I examine two linked dilemmas: the allocation of common pool water resources, and the provision of water-related public goods at multiple scales. Transaction costs can impede otherwise desired policy reforms, locking resource allocation into historic and maladaptive patterns. Transaction costs ultimately stem from multiple underlying factors shaping water allocation, including historical, geographical and technological characteristics. Understanding the nature and sources of transaction costs is therefore integral to effective policy design and institutional adaptation.

The goal of this book is to make transaction costs visible, and examine the relationship between transaction costs and institutional change in water allocation reform. I argue that the path toward sustainable and adaptive water allocation reforms involves substantial, sustained and multilevel investments in governance capacity. I elaborate a conceptual framework for analyzing the transaction costs involved in water allocation reform, which allows me to understand the factors that enhance or reduce capacity for adaptation. I draw property rights theory, transaction costs economics and common pool resource governance to understand the evolution and performance of water allocation reform, at the intersection of water markets and river basin governance. I measure and analyze transaction costs in market-oriented water rights reform and multijurisdictional water management institutions to identify factors enhancing or reducing 'adaptive efficiency' (North, 1990) – an overarching performance criterion aimed at building long-term capacity to adapt to evolving preferences and values in a context of complexity, periodic shocks, feedbacks and pervasive and deep uncertainty (see Chapter 2). I argue that water markets are not self-maintaining allocation systems (Challen, 2000) and that transaction costs help us to understand why this is the case. I show how the performance of water markets and river basin governance institutions depends on a sequence of institutional transitions that address collective action dilemmas at a range of scales – from local water rights reform to coordination institutions for basin-wide planning. Transaction costs analysis can thereby help public policy scholars to identify institutional design principles and policy lessons to inform complex water allocation trade-offs in a rapidly changing world. Through a comparative analysis of three prominent case studies, I identify design and sequencing strategies that minimize the costs of implementation, build capacity to cover necessary costs of periodic transitions, and limit the risk of decisions that lead to lock-in and unnecessarily constrain future adaptation.

AUSTRALIA AND WESTERN NORTH AMERICA AT THE LEADING EDGE

Intensified competition for scarce and variable freshwater resources have made Southeast Australia and Western North America important international examples of semi-arid transboundary rivers in federal political systems. The Colorado and Columbia Rivers of Western North America and the Murray–Darling Basin of Southeast Australia face similar water allocation challenges: competition for scarce freshwater, climate variability and change, unmet environmental needs and upstream–downstream trade-offs across subnational and, in the Columbia and Colorado, international borders. There has also been a long-standing and mutual interest in exchanging lessons between western North America and Southeast Australia. Over a century and a half of irrigation development, economic growth and diversification, and adaptation to climatic variability have made these regions laboratories for institutional reform – both successes and challenges. The environmental consequences of past development have left the needs for fish, wildlife and ecosystem services unmet.

The story of Australian Alfred Deakin's late nineteenth century visit to the US is a major milestone and entry point in the history of mutual learning between the Western US and Australia. Deakin was a major figure in Australian history in the late 1800s and early 1900s. A vigorous proponent of federation and Australia's second prime minister, he took a strong interest in irrigation development, travelling to the US and South Asia in the early 1880s for a Royal Commission study on water supply. Water supply challenges were critical for economic development and irrigation settlement in the fledgling federation. However, the political economy of water allocation and management required costly collective action to organize capital investment and manage irrigation systems, not only by private enterprise but also by irrigation districts and territorial governments across Southern Australia.

Deakin travelled to the US to learn from its experience with similar challenges of irrigation development and governance. Unlike with Europe and Asia, he considered 'the parallel between Southern Australia and the Western States of America as complete as such parallels can well be', due to 'the close resemblance of the peoples, their social and political conditions, and their natural surroundings' (Deakin, 1885, 11–12).

Notwithstanding this strong analytical and policy basis for comparison, Deakin found one major area of divergence between the Western US and Australia: the role of government versus private enterprise in irrigation development and management. In the Western US, Deakin concluded that the government had abdicated its regulatory duties, at both state and federal levels. Doing so, he believed, gave rise to multiple, often conflicting and contested, irrigation development approaches and allocation rules; a situation of institutional diversity that produced chaos rather than coherence in his estimation.

Deakin perceived state inaction as a fundamental institutional design flaw, which can now be viewed through the lens of transaction costs, property rights and coordination institutions. The Western US experience was a vivid warning to Australia of the consequences of state inaction and the corresponding risk that prohibitive transaction costs would thwart future reforms. Deakin described a 'web of litigation' that paralyzed investment and created debilitating uncertainty in the US, which remains an apt description even today. Transaction costs were not an abstract concern; for example, one irrigation canal had estimated annual litigation costs of more than £4000 (in 1880s pounds) associated with efforts to define water rights and avoid dangerous legal precedents (Deakin, 1885). Despite pockets of innovative water allocation reforms (which I discuss later in this book), fragmented institutional arrangements have persisted in the Western US. The path-dependent effect of past decisions on future reforms is the focus of Chapter 3.

Deakin's visit was a pivotal point in Australia's water development and subsequent reforms. The privatization and fragmentation of water allocation systems in the US prompted Deakin to promote a coherent policy before 'vested interests become too deeply involved' (Deakin, 1885: 21–22). Deakin strongly recommended that the state maintain control over water supply and require measurement to prevent confusion and crippling uncertainty.

The risk of tyranny by vested interests has proven more difficult than expected to combat in Australia, despite Deakin's foresight. Contemporary efforts to establish sustainable diversion limits in the Murray–Darling have involved proposals to reduce water use, triggering political resistance and stronger norms of private property rights, echoing the concerns Deakin encountered in fledgling irrigation settlements of the Western US. The experiences of the Murray–Darling and Western US have therefore remained the focus of continued comparison. In both regions, the problems of irrigation development are joined by additional pressures: sectoral competition, climate variability, environmental demands and split sovereignty within a federal political system. Transaction costs analysis can yield new insights into this connected history, recent reforms and future prospects.

My comparison of the Colorado, Columbia and Murray–Darling river basins follows the logic of most similar systems comparisons. Controlling for the many political, cultural and environmental characteristics shared by the three basins, I examine the evolution and performance of institutional design choices in market-based water rights reform and river basin governance in terms of transaction costs and adaptive efficiency, emphasizing the interaction and trade-offs across multiple performance criteria (efficiency, legitimacy, equity, sustainability) over time and space.

WATER ALLOCATION IN CLOSED BASINS

The concept of river basin closure frames the water allocation challenges in the Murray–Darling and Western US. Molle et al. (2010) define river basin closure as conditions in which downstream commitments are not met; closure is a function of societal choices about dilution, flushing and environmental flows. Basin closure highlights hydrological connections between upstream uses and downstream impacts of irrigation technology, storage infrastructure and groundwater use. Social and economic impacts of basin closure include externalities of local interventions; changes in water use patterns to meet new demands may come at the expense of existing uses or users, including environmental flows (Molle et al., 2007). In short, basin closure refers to the human causes of water scarcity as demands and development strain finite and variable water supplies.

Basin closure can be related to the economic conception of water outlined above. Water resources comprise stock (public good) and flow (common pool, private) components that are interdependent (Figure 1.2). Flow components include the renewable supplies that support human consumptive uses, and stock components include water requirements to maintain freshwater ecosystem functions and processes that sustain renewable supplies. Overextraction of surface water flows and groundwater can

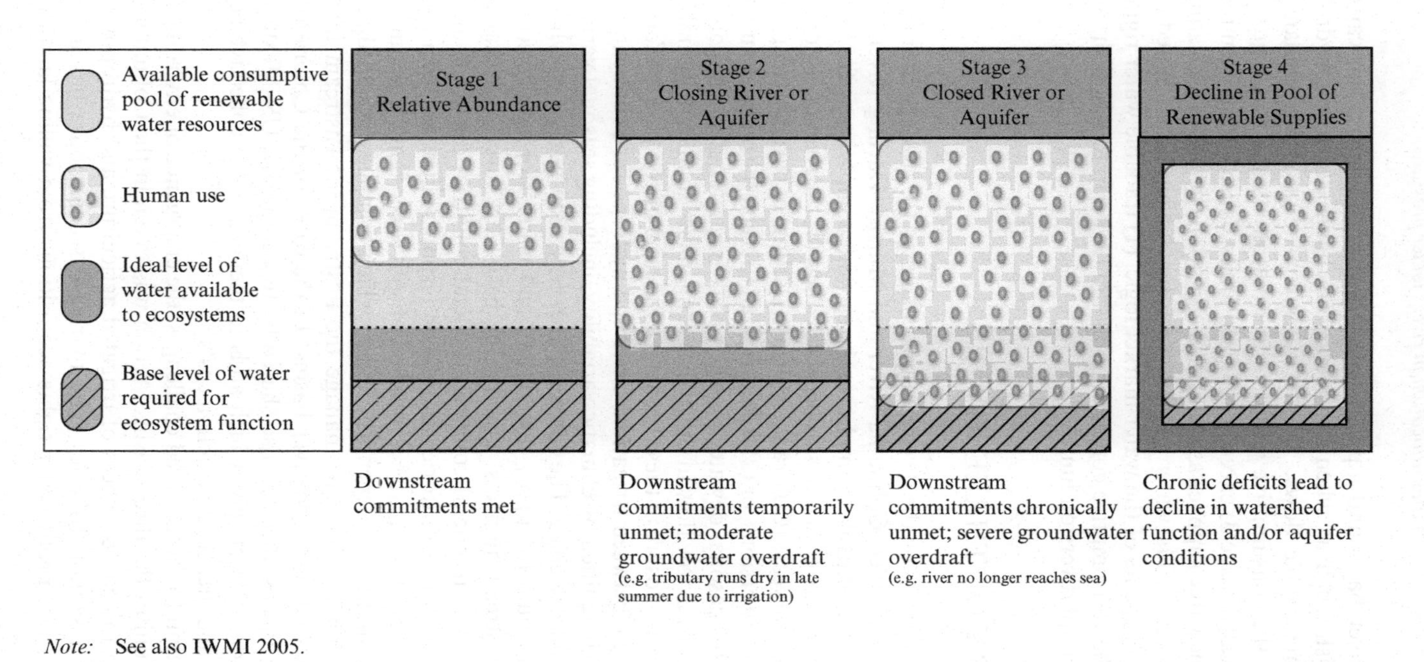

Note: See also IWMI 2005.

Figure 1.2 Stock and flow dimensions of water stress

degrade river basin and aquifer conditions and diminish their capacity to replenish, such as when subsidence or changes to forest hydrology decrease renewable flows. As the water consumption for human uses increases, the consumptive pool may periodically (stage 2 in Figure 1.2) or chronically (stage 3) exceed ecological limits, causing long-term reduction in renewable supplies as watershed conditions deteriorate or aquifers subside (stage 4). The water allocation reform challenge for a closed basin entails all elements of the water market logic: (1) tradable water rights to manage the consumptive common pool; and (2) a cap to establish, maintain and adapt diversion limits.

A STORY OF THREE RIVERS

This book focuses on the three examples of closed (or closing) river basins: the Colorado River, Columbia River and Murray–Darling River. These basins are significant examples and relevant comparisons for several reasons. They are closed basins because either: (1) the river does not reach the sea in some years (Colorado, Murray–Darling) in part due to upstream storage and diversions, or (2) tributaries fail to reach the main 'stem' of the river due to seasonal dewatering when peak irrigation demand coincides with the natural low flow in the hydrograph in late summer (Columbia).

River basin closure in these basins is a function of competition for finite water supply between irrigation, cities and hydropower production. The effects are magnified by climatic variability, including seasonal variability and drought extremes. The former makes irrigation necessary, while the latter contribute to supply uncertainty and water storage projects that have downstream impacts. Finally, the basins are large and shared by multiple states in federal countries. Federal countries distribute authority across multiple territories and divide responsibilities between federal and state levels; this creates a situation of split sovereignty with authority shared by multiple independent states, which complicates coordination at the river basin level to manage the upstream−downstream interactions prevalent in closed river basins. In the cases of the Colorado and Columbia Basins, they are also shared by two countries, US–Mexico and US–Canada, respectively, which adds to the large-scale collective action challenge and, hence, transaction costs.

In addition to these shared challenges, the emergence of environmental demands unites the three basins. Action was taken in the Columbia first to address the consequences of irrigated agriculture and hydropower on endangered salmon fisheries. The market-based response has involved public and non-profit water rights acquisitions from willing landowners (irrigation water users) to restore streamflow-dependent habitat and reconnect tributaries with main stem rivers.

THE GEOGRAPHY OF RIVER BASIN CLOSURE

Colorado River

The Colorado River (Figure 1.3a) straddles seven states in the US and two in Mexico (637 100 km^2), as well as several Indian tribes, cities and irrigation districts. It has supported extensive irrigation development (4.5 to 5.5 million acres of irrigated agriculture), hydropower production and rapid urban growth for up to 40 million people in the major population centers of the Western US (US Bureau of Reclamation, 2015). The average natural flow in the gauged record (1906–2008) at Lee Ferry is 18 500 gigalitres (GL). This average obscures high stream-flow variability: the low and high flows (6930 GL and 31 200 GL) were recorded in 1977 and 1984, respectively. Upstream reservoirs – concentrated in Lakes Powell and Mead – store up to four years of the basin's annual mean runoff to buffer against climate variability and sustained drought conditions, which are a prominent feature of the observed and paleoclimate hydrologic records (Woodhouse et al., 2006). The once vast delta ecosystem has declined due to the combination of upstream reservoirs and diversions. Projected climate change impacts include decreases in runoff, earlier snowmelt runoff and more severe droughts.

In the United States, the Colorado River Compact of 1922 apportioned water between states. Compacts are interstate apportionment mechanisms authorized by the US Congress under constitutional authority (specifically the commerce clause) and with the character of federal law (National Research Council, 2004; National Research Council, 2007). There are more than 20 in the US, and the Colorado Compact was the first. The Colorado River Compact divided the basin into two sections. The upper division states (UDS) include Colorado, New Mexico, Utah and Wyoming, and the lower division states (LDS) include Arizona, California and Nevada. The interests of each state within the UDS and LDS differ, although they will occasionally negotiate as a bloc. Mexico has a separate entitlement which was reserved in the 1922 Compact and then formalized by an international treaty in 1944. In total the 1922 Compact and 1944 Treaty commit at least 20.3 billion m^3 per year (20 300 GL) to the upper division, lower division and Mexican states. These commitments have never been fully developed due to hydrological, engineering and environmental constraints; nevertheless, water use grew steadily

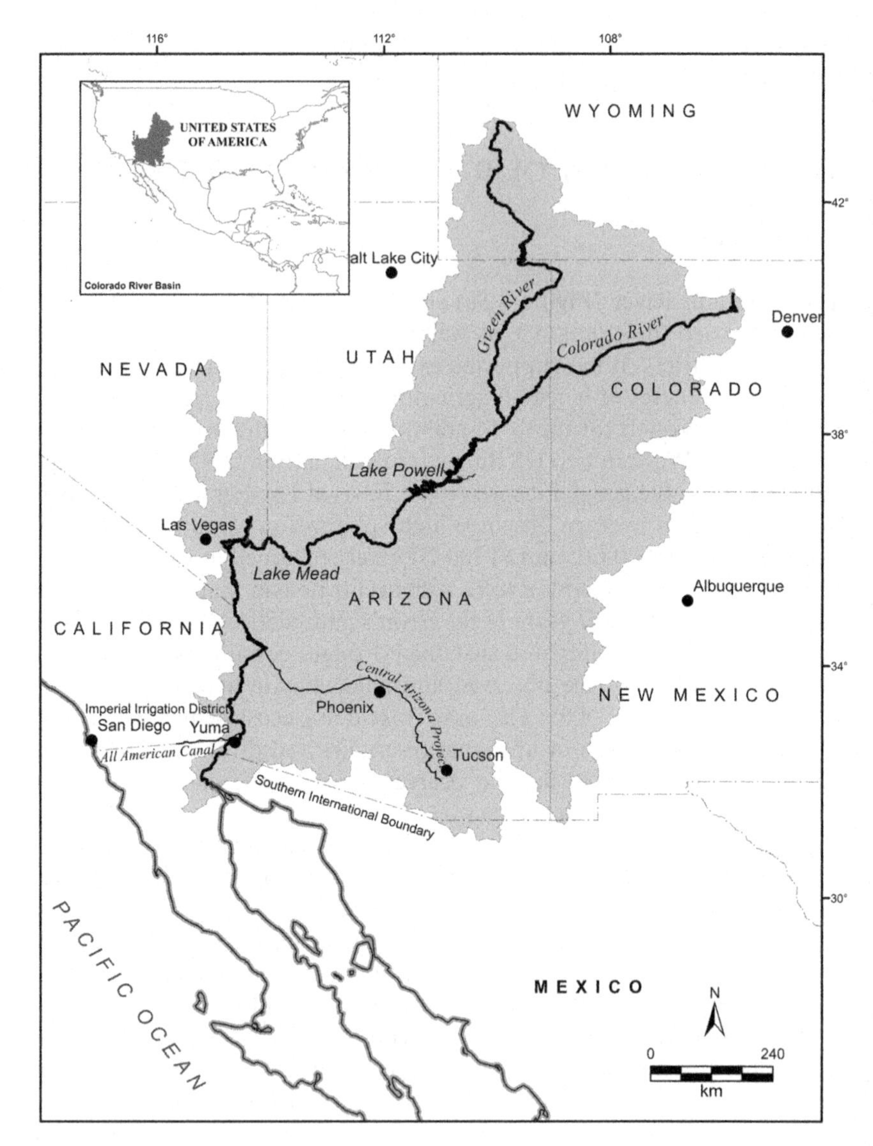

Figure 1.3a Colorado River

during the twentieth century until the early 2000s, when the Quantification Settlement Agreement required California to curb its water use to comply with its surface water entitlements, as specified by the 1963 landmark Supreme Court decision in *Arizona v. California*.

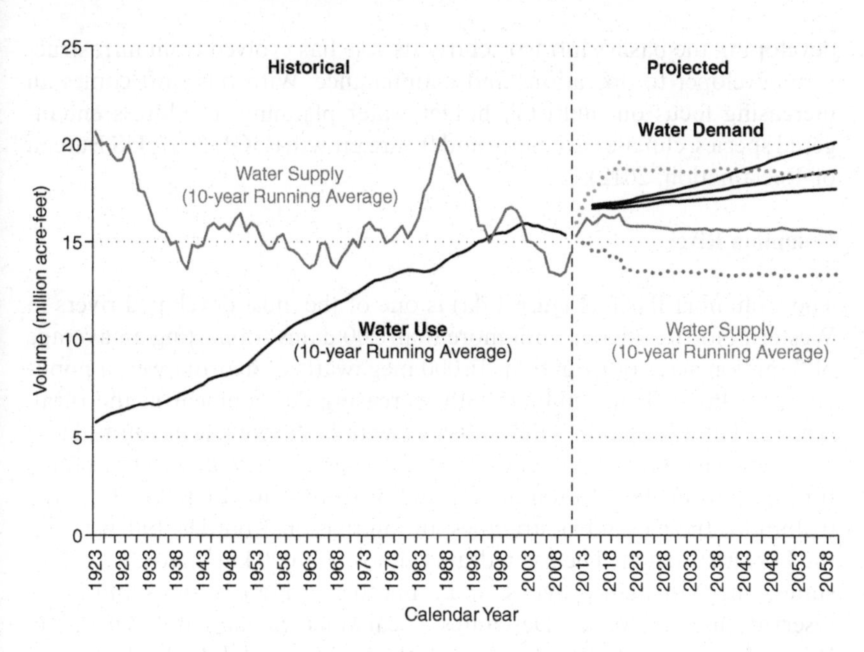

Source: US Bureau of Reclamation (2012).

Figure 1.3b River basin closure in the Colorado River Basin: long-term supply and demand trends

There is a chronic imbalance in lower basin deliveries from Lake Mead, which has been described by journalist John Fleck as a 'structural deficit' with total outflows and losses of approximately 12.6 billion m^3 and inflows of only approximately 11.1 billion m^3. The annual imbalance is buffered by reservoir storage, but this cushion has been depleted during sustained drought. Long-term demand (ten-year moving averages) intersected with long-term supply in 1999 and eclipsed it by 2002 (Figure 1.3b), depleting the storage buffers from full capacity to approximately 40 per cent as of October 2014 during an unprecedented sequence of dry years from 1999 to 2014 (US Bureau of Reclamation, 2013). These early interstate apportionment agreements are a prime example of path dependent river basin trajectories, despite a recognition of supply variability and overallocation even prior to the Compact's negotiation and certainly after the mega-drought of the 1950s. The Colorado is unique as the only major interstate river in the Western US with direct federal oversight by the Secretary of Interior, who acts as rivermaster. The US Bureau of Reclamation is the lead federal agency, derived from its historical role developing the basin's infrastructure. Its role has evolved from infrastructure developer to operations and maintenance. With this shift comes an increasing focus on multistakeholder water planning to address endangered species, climate variability and future growth (DOI, 2007; US Bureau of Reclamation, 2012).

Columbia River

The Columbia Basin (Figure 1.4a) is one of the most developed rivers in Western North America with more than 200 dams supporting an average of 3 million acres of irrigation; 16 000 megawatts of hydropower; a population of more than 7 million with increasing development in the rural, semi-arid interior; and a salmon fishery with high ecological, cultural and economic significance. A series of 31 dams generate hydroelectric power through the Federal Columbia River Power System (FCRPS). Like the Colorado, the Columbia straddles an international border, but between the US and Canada. The basin drains almost 700 000 km^2 across seven US states, one Canadian province and a number of First Nations and tribal reservations. The Snake, Deschutes, Clearwater, Salmon and Willamette Rivers form important tributaries to the mainstem, while the John Day remains the longest free-flowing river in the basin (that is, without dams).

The Columbia River Basin is comparable in size with the Colorado in terms of drainage area but not in volume. The river discharges into the Pacific Ocean west of Portland, Oregon, after descending 820 m over 2000 km from its headwaters in British Columbia, Canada. It has an average volume at the Dalles Dam of 165 billion m^3, an order of magnitude higher than the Colorado (NPCC, 2014). The annual average inflows make the Columbia the fourth largest river in North America by volume (National Research Council, 2004). However, like the Colorado and Murray–Darling, stream flow is characterized by spatial and seasonal variability due to a snowmelt dominated hydrograph. Prior to river basin development, 75 per cent of annual runoff occurred during the late spring and summer and only 25 per cent during the fall and winter. This seasonal variability has led to efforts to develop the river through the FCRPS, which has transformed the river's hydrology (flattening out seasonal variability), geomorphology and ecology. Tributaries upstream of these reservoirs still experience seasonal variability. Chronic seasonal water deficits in the tributaries occur in late summer (August, September), when peak agricultural use coincides with natural low flows after snowmelt (Figure 1.4b).

The Columbia is managed by a 'patchwork quilt' of laws, policies and jurisdictions, with allocation authority vested at the state level (Schoessler

Figure 1.4a Columbia River and selected tributaries

et al., 1997). The Columbia differs from the Colorado because it lacks a lead federal agency in water planning; it also lacks an interstate compact to apportion water rights between the upper and lower basin states, despite an attempt from 1950 to 1968 focused principally on coordinating

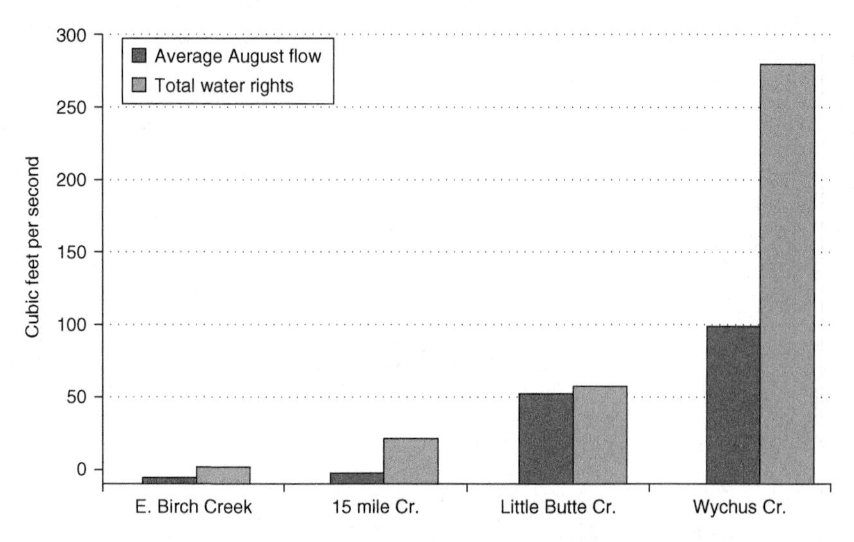

Figure 1.4b River basin closure in the tributaries of the Columbia: average August streamflows in selected tributaries of the Columbia River, compared with total water rights

hydropower generation between private and public entities (National Research Council, 2004). The FCRPS establishes a federal overlay in water management for power and conservation activities under the 1980 Northwest Power Act. Tribes also have substantial property rights, including 1855 Stevens Treaty rights to salmon harvests (Bark et al., 2012). The 1964 international treaty shares flood control and hydropower benefits between Canada and the United States, but is silent on water quantity, ecosystem protection and other matters. An interstate apportionment treaty has been elusive in the US portion of the Columbia Basin. Federal energy and environmental laws have facilitated interstate coordination through the 1980 Northwest Power Act and the authorization of the Bonneville Power Administration as a quasi-governmental energy utility. The region-wide scope of both entities influences water-related planning, salmon recovery and infrastructure management.

Murray–Darling

The Murray–Darling Basin (Figure 1.5a) is more than 1 million km^2 and 14 per cent of Australia's landmass, spanning four states and one territory. The Great Dividing Range forms the eastern boundary and feeds the Murray and Murrumbidgee Rivers. The Darling River of

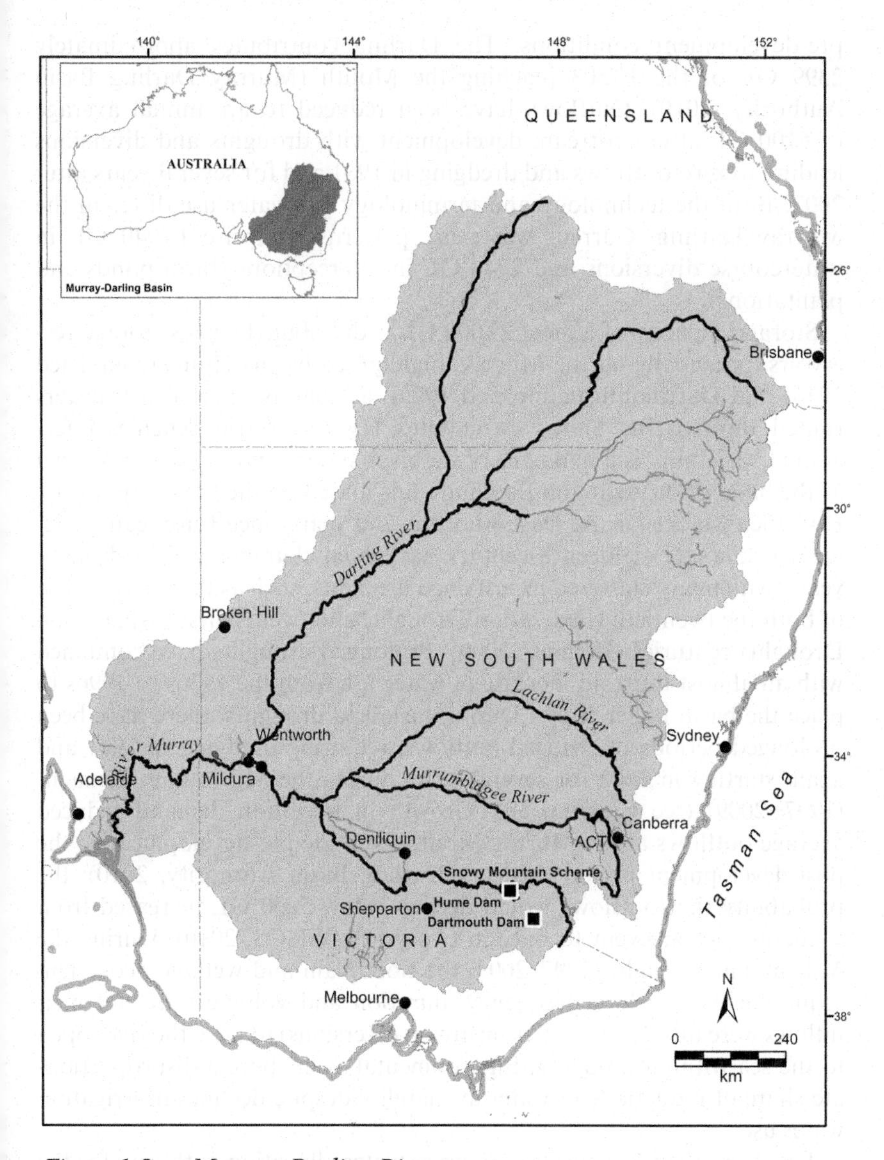

Figure 1.5a Murray–Darling River

Southern Queensland and New South Wales is influenced by a different set of hydroclimatic influences with higher temperatures and rainfall variability than the Southern Murray system (CSIRO, 2008). The three rivers are the longest in Australia, with an inflow of 32 800 GL and an annual average outflow of 12 500 GL to the Murray Mouth under pre-development conditions. The Darling contributes approximately 2399 GL of the 12503 reaching the Mouth (Murray–Darling Basin Authority, 2010). Outflows have been reduced to an annual average of 5100 GL after upstream development with droughts and diversions leading to zero outflows and dredging in 1981 and for several years after 2002. Both the technology and terminology for water use differ in the Murray–Darling. Current water use patterns comprise 10940 GL in watercourse diversions and 2740 GL in interceptions (farm ponds and plantations).

Storage capacity of almost 22000 GL is distributed across storage reservoirs, principally on the Murray, highlighted by the Hume (completed 1936) and Dartmouth (completed 1979), as well as interbasin transfers routed through the Snowy Mountains Hydro-Electric Scheme. Interannual variability is a hallmark of the river system and the region known as the 'land of drought and flooding rains' based on the famous poem by Dorothea Mackellar. At least 80 of the 240 years since European settlement in the late eighteenth century have been characterized as drought years, with many clustered in sustained droughts, such as those at the turn of both the twentieth (Federation Drought) and twenty-first (Millennium Drought) centuries (Helman, 2009). Prolonged droughts have combined with an almost fourfold increase in water use from the 1930s to 1990s to place the basin under stress. During sustained droughts, there have been prolonged periods of reduced outflows, requiring dredging in 1981 and again starting in 2002 for several years during the Millennium Drought (1997−2009) (see Figure 1.5b). Growth in irrigation demand reduced average outflows at the Murray Mouth from the pre-development to the post-development period (Murray–Darling Basin Authority, 2010); the probability that outflows would decline below 5000 GL increased from a one in 20-year event to once in two years (WGCS, 2010). During the Millennium Drought (1997–2009) the floodplain and wetland ecosystem requirements for flood frequency, duration and volume were not met; inflows were insufficient (after upstream diversions) to keep the river open to the sea without dredging. Environmental assets bore a disproportionate share of reductions in water availability despite declines in irrigation water use.

The Australian Constitution reserves water allocation authority for the states, which allocate water via statutory water plans. The 1914–15 River Murray Waters Agreement divided water among the three main states in the Southern Murray: Victoria, New South Wales and South Australia. The Murray–Darling Basin Agreement (1992) expanded cooperation to Queensland and the Australian Capital Territory. The Council of Australian Governments adopted a framework for reform in 1994, which

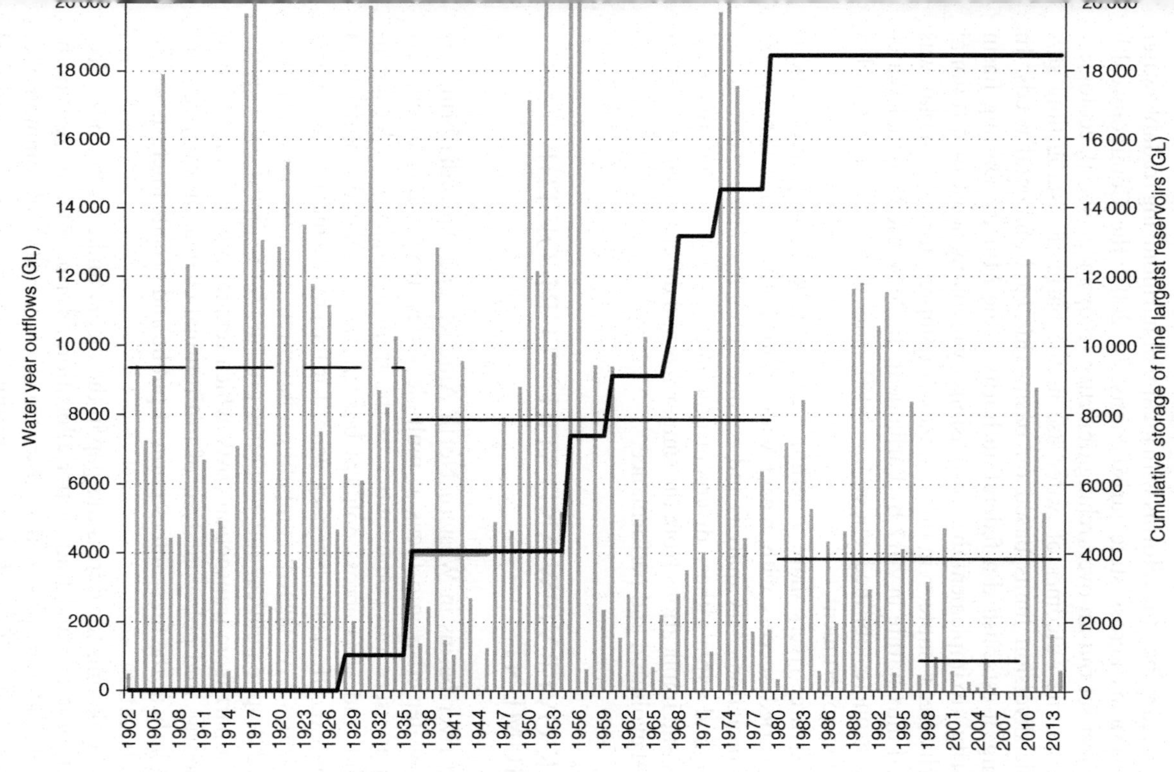

Note: Water year outflows from the Murray Mouth. Horizontal lines represent averages for the given periods before, during and after storage development, as well as during the Millennium Drought. Outflows eclipsed 20000 GL (and are off the chart) in 1917, 1952, 1955, 1956 and 1974. Stepped curve represents the cumulative storage of the nine largest reservoirs, as identified in the 2014 Annual Report of the Murray Darling Basin Authority.

Source: Data from Murray–Darling Basin Authority, includes modelled outflows for the early 1900s.

Figure 1.5b River basin closure in the Murray–Darling Basin

led to the subsequent audit of water use and a cap on diversions in 1995. The 2004 National Water Initiative (NWI) – an intergovernmental agreement among four states, one territory and the Commonwealth government – streamlined the reforms with an aim of harmonizing tradable water rights systems across state jurisdictions. In 2007, the Commonwealth Water Act set out an overarching objective to optimize social, ecological and economic outcomes of water use in the basin by establishing sustainable diversion limits based on basin-wide environmental needs. In so doing, it established a federal authority – the Murray–Darling Basin Authority – to implement these provisions by invoking new Constitutional authority (see Chapter 5 in this volume). A comprehensive basin plan was adopted into statute in 2012 by Parliament. In 2014, the National Water Commission (established under the NWI) was disbanded as part of a change in government.

These three rivers share a defining set of challenges but have experienced varied success with recent water allocation reforms. This book uses transaction costs analysis and common pool resource governance to explore and explain these divergent outcomes, as well as the principles, lessons and implications from the similarities and differences in their institutional evolution and performance.

BOOK OVERVIEW AND STRUCTURE OF THE ARGUMENT

Closed river basins in Western North America and Australia bring the global water crisis into focus at multiple scales, from the local impacts on water users and communities to the transboundary interdependencies and trade-offs across political borders from states to countries. In this context, efforts to claw back water for the environment over the past two decades have lagged, competition has intensified, and conflicts have lingered despite political and economic pressure for reallocation. This lag stems in part from transaction costs, which remain poorly understood in our theories and policy analysis of institutional change in water reform, particularly at the fuzzy intersection of public and private interests within water allocation.

The book aims to address this gap in the theory and evidence on collective action in cap-and-trade water allocation reforms, by linking conceptual and analytical perspectives from transaction costs and common pool resource studies. It will do so by applying conceptual advances, empirical evidence and methodological innovations to three case studies of market-based water reform with varying levels of success.

The objectives of the book are addressed as follows. Chapter 2 introduces an analytical framework for examining the evolution and performance of water allocation institutions. I synthesize concepts, evidence and methodological approaches in transaction costs analysis by: framing the challenge of adaptation as a challenge of reallocation, introducing the transaction as a unit of analysis, and examining complex, mixed property rights for common pool resources. I draw on four theoretical traditions in transaction costs and institutional analysis: Coase, Ostrom, Williamson, and North, as well as several important property rights economists who have focused on water (Anderson and Libecap to name a couple). A key insight lies in the concept of path dependency: the fact that past decisions enable some options and foreclose others, demonstrating the importance of interactions between the costs of implementing transactions within the prevailing institutional setting versus the costs of institutional transitions to water rights systems and river basin governance institutions. In this context, I build on recent efforts to elaborate adaptive efficiency as a multi-dimensional performance criterion for assessing the evolution and performance of water allocation reforms in a transaction costs world.

Chapter 3 traces the path dependence of water resource development and institutional reform across the three river basins to frame the past trajectories, current trade-offs and future prospects of water allocation reforms. It highlights three important institutional choices and technological innovations associated with water rights systems, irrigation supply organizations and interstate apportionment agreements. It shows that the assumptions guiding initial water allocation decisions no longer appear valid: the notion of a 'stationary' water supply, the predominance of irrigation and the acceptable level of impacts on the environment. The notion of 'lock-in' costs is used to understand the institutional friction of transaction costs and the intertemporal trade-offs of path dependent policy reforms in the early phases of cap-and-trade water allocation reforms across the linked processes of capping, allocating and trading. The Colorado River Basin demonstrates the lock-in consequences of past decisions; allocation decisions taken in the early twentieth century were based on optimistic supply scenarios that guided water resource development and created unexpected (downstream) winners and (upstream and environmental) losers. This lock-in is a starting point to consider reform lessons from promising examples in the Columbia and Murray–Darling Basins, as well as recent breakthroughs in the Colorado itself.

Chapter 4 analyzes the emergence, evolution and performance of markets for environmental flows in the Columbia Basin. It measures and explains transaction costs and adaptive efficiency across 13 watershed-level case studies in four US states over an eight-year period from 2003–2010. The performance of emerging markets for environmental flows varies as much within states as across them, due to importance of local institutions for water rights reform and the multilevel collective action at the field, state and federal levels to coordinate market transactions with wider institutional frameworks for water planning, water banking and so on. Empirical analysis of water reallocation for environmental flows in the Columbia Basin is used to demonstrate these conceptual issues and provide methodological tools to measure and manage the transaction costs of water rights reform. Across the three basins, diversion limits have been based on historic use patterns, which engrained unsustainable extraction levels in property rights institutions.

Chapter 5 examines the maturation of markets and provision of multiple water-related public goods in the Murray–Darling, each with different communities of interest, politics and economies of scale. I examine efforts to 'scale up' water allocation reform across three interdependent elements: market-based water rights reform, the establishment (and adaptation) of diversion limits, and the recovery of water for the environment to address the consequences of overallocation. The challenge of scaling up trading activity and recovering water for the environment exposes the need for substantial, sustained and multi-level governance capacity. This capacity depends on a mixture of formal and informal institutions coordinated by polycentric governance arrangements, highlighting the importance of institutional mechanisms to balance subsidiarity (the local capacity necessary in the Columbia example) with complementarity (the diverse vertical and horizontal coordination mechanisms to reconcile tasks and trade-offs that span jurisdictions and interests).

The final chapter reflects on the lessons for the theory and practice of water allocation reform in rivers under pressure, with implications for future studies to diagnose governance challenges, and design and sequence the reforms needed for adaptive efficiency.

NOTES

1. Although the US has been able to decouple population growth and water use (Gleick, 2003).
2. Ostrom et al. focus on 'global commons' but also refer to large international rivers as an example of large-scale common pool resources.
3. These correspond roughly to command-and-control regulation, private property and community-based management, respectively.
4. IWRM is a process which promotes the coordinated development and management of

water, land and related resources, in order to maximize the resultant economic and social welfare in an equitable manner without compromising the sustainability of vital ecosystems (Global Water Partnership, 2000).

5. See Chapter 2 for a thorough consideration of rivalling definitions of transaction costs, ranging from broad and dynamic to narrow and static.

REFERENCES

ANDERSON, T.L. and LEAL, D.R. 2001. *Free Market Environmentalism*. New York: Palgrave Macmillan.

ANDERSON, T.L., SCARBOROUGH, B. and WATSON, L.R. 2012. *Tapping Water Markets*. London: Routledge.

ANDERSSON, K.P. and OSTROM, E. 2008. Analyzing decentralized resource regimes from a polycentric perspective. *Policy Sciences*, 41, 71–93.

BAKKER, K. 2014. The business of water: market environmentalism in the water sector. *Annual Review of Environment and Resources*, 39, 469–94.

BARK, R. H., GARRICK, D. E., ROBINSON, C. J. and JACKSON, S. 2012. Adaptive basin governance and the prospects for meeting Indigenous water claims. Environmental Science & Policy, 19, 169–177.

BENNETT, J. 2005. *The Evolution of Markets for Water: Theory and Practice in Australia*. Edward Elgar Publishing, Cheltenham, UK.

BLOMQUIST, W. and SCHLAGER, E. 2005. Political pitfalls of integrated watershed management. *Society and Natural Resources*, 18, 101–17.

BLUMM, M.C. 1992. Fallacies of free market environmentalism. *Harvard Journal of Law and Public Policy*, 15, 371.

CAREY, J.M. and SUNDING, D.L. 2001. Emerging markets in water: a comparative institutional analysis of the Central Valley and Colorado–Big Thompson projects. *Natural Resources Journal*, 41, 283–328.

CHALLEN, R. 2000. *Institutions, Transaction Costs and Environmental Policy*. Cheltenham, UK and Northampton, MA, USA: Edward Elgar Publishing.

CHOPRA, K.R. and MILLENNIUM ECOSYSTEM ASSESSMENT RESPONSES WORKING GROUP. 2005. *Ecosystems and Human Well-Being: Policy Responses: Findings of the Responses Working Group*. Washington, DC: Island Press.

COMMONWEALTH SCIENTIFIC AND INDUSTRIAL RESEARCH ORGANISATION (CSIRO). 2008. *Water Availability in the Murray–Darling Basin Report*. Canberra: CSIRO Publishing.

CONCA, K. 2006. *Governing Water: Contentious Transnational Politics and Global Institution Building*. Cambridge, MA: MIT Press.

CONGRESSIONAL RESEARCH SERVICE. 2009. 35 years of water policy: the 1973 National Water Commission and present challenges. In: CODY, B.A. and CARTER, N.T. (eds). Washington, DC.

COX, M., ARNOLD, G. and TOMÁS, S.V. 2010. A review of design principles for community-based natural resource management. *Ecology & Society*, 15(4), 38.

DEAKIN, A. 1885. Royal Commission on Water Supply, first progress report, irrigation in Western America. Available at: http://www.dwu.edu/library/

environmental_history/ElectronicArchivesPDFDocuments/WaterConservation&Development/deakinroyalcommissiononwatersuupply.pdf.

EASTER, W.K., ROSEGRANT, M.W. and DINAR, A. 1998. *Markets for Water: Potential and Performance*. Boston: Kluwer Academic Publisher.

EASTER, K.W. AND HUANG, Q. 2014. *Water Markets for the 21st Century: What Have We Learned?* Dordrecht: Springer.

GARRICK, D., WHITTEN, S. and COGGAN, A. 2013. Understanding the evolution and performance of water markets and allocation policy: a transaction costs analysis framework. *Ecological Economics*, 82, 195−205.

GLEICK, P.H. 2003. Global freshwater resources: soft-path solutions for the 21st Century. *Science*, 302, 28 November, 1524–28.

GLENNON, R.J. 2010. *Unquenchable: America's Water Crisis and What To Do About It*. Washington, DC: Island Press.

GRAFTON, R.Q. and HORNE, J. 2014. Water markets in the Murray–Darling Basin. *Agricultural Water Management*, 145 (2014), 61–71.

GRAFTON, R.Q., LIBECAP, G., MCGLENNON, S., LANDRY, C. and O'BRIEN, B. 2011. An integrated assessment of water markets: a cross-country comparison. *Review of Environmental Economics and Policy*, 5 (2), 219–39.

GRAFTON, R.Q., LIBECAP, G.D., EDWARDS, E.C., O'BRIEN, R.J. and LANDRY, C. 2012. Comparative assessment of water markets: insights from the Murray–Darling Basin of Australia and the Western USA. *Water Policy*, 14, 175.

HAMLET, A.F. 2003. The role of transboundary agreements in the Columbia River Basin: an integrated assessment in the context of historic development, climate, and evolving water policy. *Climate and Water*, 16, 263−89.

HANAK, E. and STRYJEWSKI, E. 2012. California's water market, by the numbers: update 2012. Public Policy Institute of California.

HELMAN, P. 2009. Droughts in the Murray Darling Basin since European Settlement. Griffith Centre for Coastal Management Research Report No. 100. Canberra: Murray−Darling Basin Authority.

HOWE, C.W., BOGGS, C. and BUTLER, P. 1990. Transactions costs as determinants of water transfers. *University of Colorado Law Review*, 61(2), 393–405.

INTERNATIONAL CONFERENCE ON WATER AND THE ENVIRONMENT (ICWE). 1992. *The Dublin Statement on Water and Sustainable Development*. Dublin: World Meteorological Organization. Available at: http://www.wmo.int/pages/prog/hwrp/documents/english/icwedece.html.

INTERNATIONAL WATER MANAGEMENT INSTITUTE. 2005. Planning for Environmental Flows. Water Policy Issue Paper 15. IWMI, Sri Lanka. Available at: http://www.iwmi.cgiar.org/Publications/Water_Policy_Briefs/PDF/wpb15.pdf.

MAESTU, J. 2013. *Water Trading and Global Water Scarcity: International Experiences*. Washington, DC: RFF Press.

MARSHALL, G. 2005. *Economics for Collaborative Environmental Management: Renegotiating the Commons*. Abingdon: Routledge.

MARSHALL, G. 2008. Nesting, subsidiarity, and community-based environmental governance beyond the local scale. *International Journal of the Commons*, 2, 75−97.

MCGINNIS, M.D. and OSTROM, E. 2012. Reflections on Vincent Ostrom, public administration, and polycentricity. *Public Administration Review*, 72, 15−25.

MEINZEN-DICK, R. 2007. Beyond panaceas in water institutions. *Proceedings of the National Academy of Sciences*, 104, 15200–15205.
MEINZEN-DICK, R. 2014. Property rights and sustainable irrigation: A developing country perspective. *Agricultural Water Management*.
MOLLE, F. 2008. Nirvana concepts, narratives and policy models: insights from the water sector. *Water Alternatives*, 1, 131–56.
MOLLE, F. 2009. River-basin planning and management: the social life of a concept. *Geoforum*, 40, 484–94.
MOLLE, F., WESTER, P. and HIRSH, P. 2010. River basin closure: processes, implications and responses. *Agricultural Water Management*, 97, 569–77.
MOSS, T. 2004. The governance of land use in river basins: prospects for overcoming problems of institutional interplay with the EU Water Framework Directive. *Land Use Policy*, 21, 85–94.
MOSTERT, E., CRAPS, M. and PAHL-WOSTL, C. 2008. Social learning: the key to integrated water resources management? *Water International*, 33, 293–304.
MURRAY–DARLING BASIN AUTHORITY (MDBA). 2010. *Guide to the Proposed Basin Plan*. Canberra: Murray–Darling Basin Authority.
NATIONAL RESEARCH COUNCIL. 1992. Water Transfers in the West: Efficiency, Equity, and the Environment. Washington, DC: National Academies Press.
NATIONAL RESEARCH COUNCIL. 2004. Managing the Columbia River Instream Flows, Water Withdrawals, and Salmon Survival.
NATIONAL RESEARCH COUNCIL. 2007. Colorado River Basin Water Management Evaluating and Adjusting to Hydroclimatic Variability. Washington, D.C: NRC.
NATIONAL WATER COMMISSION (NWC). 1973. *Water Policies for the Future: Final Report to the President and to the Congress of the United States by the National Water Commission*. Washington, DC: National Water Commission.
NEUMAN, J.C. 2004. The good, the bad, and the ugly: the first ten years of the Oregon Water Trust. *Nebraska Law Review*, 83, 432.
NORTHWEST POWER AND CONSERVATION COUNCIL (NPCC). 2014. Fiscal Year 2013 Annual Report: The State of the Columbia River Basin. NPCC, Portland. Available at: https://www.nwcouncil.org/reports/financial-reports/2014-01/.
OECD. 2011. *Water Governance in OECD Countries: A Multi-Level Approach*. Paris: OECD Publishing.
OECD. 2012. *OECD Environmental Outlook to 2050*. Paris: OECD Publishing.
OLMSTEAD, S.M. 2010. The economics of managing scarce water resources. *Review of Environmental Economics and Policy*, 4(2), 179–98.
OSTROM, E. 1990. *Governing the Commons: The Evolution of Institutions for Collective Action*. Cambridge: Cambridge University Press.
OSTROM, E., BURGER, J., FIELD, CB., NORGAARD, RB. and POLICANSKY, D. 1999. Revisiting the commons: local lessons, global challenges. *Science*, 284 (5412), 278–282
OSTROM, E. 2009. *Understanding Institutional Diversity*. Princeton, NJ: Princeton University Press.
OSTROM, V., TIEBOUT, C.M. and WARREN, R. 1961. The organization of government in metropolitan areas: a theoretical inquiry. *American Political Science Review*, 55, 831–42.

RITTEL, H.W. and WEBBER, M.M. 1973. 2.3 planning problems are wicked. *Polity*, 4, 155–69.
ROSE, C.M. 1990. Energy and efficiency in the realignment of common-law water rights. *Journal of Legal Studies*, 261–96.
ROSEGRANT, M.W. and BINSWANGER, H.P. 1994. Markets in tradable water rights: potential for efficiency gains in developing country water resource allocation. *World Development*, 22(11), 1613–25.
SALIBA, B. and BUSH, D.B. 1987. *Water Markets in Theory and Practice: Market Transfers, Water Values, and Public Policy*. Westview Press.
SCHEWE, J., HEINKE, J., GERTEN, D., HADDELAND, I., ARNELL, N.W., CLARK, D.B. . . . and KABAT, P. 2014. Multimodel assessment of water scarcity under climate change. *Proceedings of the National Academy of Sciences*, 111(9), 3245–50. Available at: http://www.pnas.org/content/111/9/3245.full.pdfMar.
SCHLAGER, E. and BLOMQUIST, W.A. 2008. *Embracing Watershed Politics*. Boulder, CO: University Press of Colorado.
SCHLAGER, E. and OSTROM, E. 1992. Property-rights regimes and natural resources: a conceptual analysis. *Land Economics*, 68, 249–62.
SCHOESSLER, M.A., BLUMM, M.C. and SWIFT, B.M. 1997. *A Survey of Columbia River Basin Water Law Institutions and Policies: Report to the Western Water Policy Review Advisory Commission*. Denver, CO: Western Water Policy Review Advisory Commission.
THIEL, A. 2014. Rescaling of resource governance as institutional change: explaining the transformation of water governance in southern Spain. *Environmental Policy and Governance*, 24, 289–306.
US BUREAU OF RECLAMATION. 2012. *Colorado River Basin Water Supply and Demand Study*. Washington, DC: Department of Interior.
US BUREAU OF RECLAMATION. 2013. *Upper Colorado Region Water Resources Group Annual Operating Plans*. Washington, DC: Department of Interior.
US BUREAU OF RECLAMATION. 2015. *Moving Forward to Address the Challenges Identified in the Colorado River. Basin Water Supply and Demand Study*. Washington, DC: Department of Interior.
US DEPARTMENT OF THE INTERIOR (DOI). 2007. Record of decision: Colorado River interim guidelines for lower basin shortages and the coordinated operations for Lake Powell and Lake Mead.
VOROSMARTY, C.J., MCINTYRE, P.B., GESSNER, M.O., DUDGEON, D., PRUSEVICH, A., GREEN, P., GLIDDEN, S., BUNN, S.E., SULLIVAN, C.A., REIDY LIERMANN, C. and DAVIES, P.M. 2010. Global threats to human water security and river biodiversity. *Nature*, 467, 555–61.
WENTWORTH GROUP OF CONCERNED SCIENTISTS (WGCS). 2010. Sustainable Diversions in the Murray–Darling Basin: An analysis of the options for achieving a sustainable diversion limit in the Murray–Darling Basin. Canberra, Australia.
WOODHOUSE, C.A., GRAY, S.T. and MEKO, D.M. 2006. Updated streamflow reconstructions for the Upper Colorado River basin. *Water Resources Research*, 42, 1–16.
WORLD ECONOMIC FORUM (WEF). 2013. *Global Risks 2013*, 8th edition. Davos, Switzerland.
WORLD ECONOMIC FORUM (WEF). 2015. *Global Risks 2015*, 10th edition, Davos, Switzerland.

2. Water allocation and institutional change in a transaction costs world: an analytical framework

INTRODUCTION

In this chapter, I frame the challenge of water allocation in a world of positive and often substantial transaction costs. Multiple theoretical traditions have examined the relationship between transaction costs and institutional change in natural resource allocation, including four Nobel laureates in the last 25 years: Coase, Williamson, Ostrom and North. Although each offers an important perspective, none is sufficient on its own to understand the evolution and performance of water allocation policy. Blending perspectives from Coase, Williamson, North and Ostrom (C-WON), allows the analyst to embrace complexity and elaborate a new calculus of institutional change for individuals and a 'winning coalition' needed for market-oriented water rights reform and river basin governance. This analytical perspective uses transactions as units of analysis, establishes models and typologies of transaction costs and institutional change, and enables the measurement and evaluation of institutional reforms in terms of adaptive efficiency, requiring ongoing institutional transitions to solve wicked water allocation problems in a path dependent world.

WATER ALLOCATION IN CLOSED RIVERS: A CHALLENGE OF REALLOCATION AND ADAPTATION

Water allocation reform in closed river basins is often a challenge of reallocation. Meeting new demands for freshwater resources in fully appropriated rivers may require a change in water allocation. When rivers no longer reach the sea, emerging demands are either unmet, or they are satisfied through supply augmentation, demand conservation or reallocation from existing consumptive users. Increasingly, the three options are pursued simultaneously as part of a portfolio approach. The ability to reallocate water also provides incentives for water conservation and disincentives for costly new supply infrastructure when reallocation offers a lower-cost alternative.

The options for reallocation vary along two dimensions: whether they are voluntary and whether they are compensated. Market-based water allocation mechanisms enable voluntary reallocation based on a transaction between a willing buyer and willing seller, ostensibly based on a price reflecting the 'scarcity value' of the water, that is, the benefits forgone by keeping the water in its present use. Administrative reallocation is involuntary and can either be compensated or not depending on the protections against regulatory changes that impair legally protected ownership interests (Colby, 1988). Water allocation institutions establish cumulative limits on diversions, systems of water rights, and rules governing reallocation. These water institutions seek to match supply and demand according to multiple, and often competing, criteria of fairness, economic efficiency, accountability, robustness, or, most likely, some combination of these (Ostrom, 2009a; Howe, 2000). Reallocation is integral to adaptation, and it hinges on institutional change to establish or reform tradable water rights systems, manage associated trade-offs and third party equity concerns, and coordinate across political jurisdictions when defining or adapting the cumulative limits on diversions.

Reallocation is not guaranteed to happen even when there is a strong economic justification and political will. Transaction costs have been described as 'the economic equivalent of friction' (Williamson, 1985: 2) and can constrain reallocation and wider efforts to create or modify property rights institutions. As such, they constitute an important influence on institutional change and economic performance, biasing changes in property rights and resource allocation to incremental changes (North, 2006, 1994, 1990). While efficiency requires the reduction of transaction costs to ensure the least-cost[1] path to a given policy target, other policy objectives impose transaction costs (for example, seeking equity, fairness, robustness). Therefore, transaction costs play an important role in maintaining order and political legitimacy in property rights systems by balancing stability and flexibility in a dynamic tension (Bromley, 1989).

Voluntary water reallocation has occurred in all three basins considered in this volume. Trading activity in the three river basins – albeit with varying frequencies and magnitudes – is a tacit sign that fundamental institutional barriers have been addressed and that the benefits of doing so have increased as competition and scarcity intensify. In a transaction costs world, this is often achieved with great political and economic difficulty after institutional transitions requiring multiple decades. Even then many of the potential gains from trade may not be realized. This process of institutional change can be accelerated by droughts or shortages which expose the potential benefits (increased scarcity value of water and differentials in water value across uses) and expand the number of potential beneficiaries of trading. This can be accompanied by efforts, even temporary, to reduce the administrative barriers. This is not guaranteed, however, as illustrated by maladaptations during drought which may provide temporary benefits but decrease long-term resilience and adaptive efficiency (Christian-Smith et al., 2014).

Water allocation reforms in the stressed rivers of Western North America and Southeast Australia seek to address the linked dilemmas of allocating common pool water resources and providing multiple water-related public goods. Assessing such reforms requires attention to the efficiency of water transactions but must also be considered in terms of the other criteria needed for long-term adaptability, equity and legitimacy. The performance of water allocation reforms is a function of transaction costs and property rights in the context of prevailing social, physical, technological and political economic circumstances, including production or transformation costs. Transaction costs analysis provides a lens for understanding and assessing the evolution and performance of these reforms.

This chapter sets out the conceptual pillars, analytical framework and empirical evidence about the relationship between transaction costs and institutional change in market-based water allocation policy. It uses the transaction as the unit of analysis, and examines a set of linked transactions occurring within water markets and political borders nested from irrigation districts to international boundaries. Definitions matter, and the term 'transaction costs' has become the subject of a sprawling literature, with increasingly expansive definitions since Coase introduced the concept (Coase, 1960, 1988; McCann et al., 2005; McCann and Easter, 2004; Krutilla and Krause, 2010; Marshall, 2013). The point of departure is a broad view of transaction costs as comprising the resources (money, time, staff, and so on) incurred in collective action to define, manage and transfer property rights to water, create and change institutional arrangements, and organize the provision of multiple, often competing, public goods (Garrick et al., 2013a; McCann et al., 2005; McCann and Easter, 2004; Marshall, 2013; Hagedorn, 2008). This broad definition can in turn be decomposed into subtypes and chronologies of transaction costs, each subject to different drivers and interactions. Models of transaction costs and natural resource governance can be framed in terms of 'calculus' of institutional change; institutional change is expected when the benefits of a rule change are sufficient to overcome the transaction costs, both the up-front and ongoing transaction costs incurred, with a given level of technological innovation and transformation costs. Adaptive efficiency offers a performance criterion to understand and evaluate water allocation reform in rivers under pressure; it can only be developed and assessed over the long-term. The analytical framework offers conceptual distinctions to operationalize adaptive efficiency across multiple, interacting performance variables over space and time. The next sections elaborate the conceptual pillars, analytical framework and performance criteria used to learn from the reform experiences in the three rivers.

CONCEPTUAL PILLARS

The Transaction: From Unit of Analysis to Linked Action Arenas

The transaction is the unit of analysis for water reallocation. The transaction connects potential buyers, sellers, regulators and a range of affected stakeholders who cooperate and compete in freshwater development, access, withdrawal and management. In the first half of the twentieth century, John Commons identified the transaction as the unit of analysis for understanding collective action in 'control, liberation or expansion' of individual behavior. Transactions involve multiple actors whose interactions are governed by a set of 'working rules' (Commons, 1931). In the context of water, working rules form the property institutions governing water allocation, conflict resolution and coordination to provide water-related public goods. Commons argued that such institutions for collective action are based on interdependence and the need for order, which applies to socially and physically interconnected water resources.

The transaction provides the entry point into a transaction costs world – a 'world of collective action and perpetual change, which is the uncertain future world' (Commons, 1931). The transaction is an example of an 'action arena' in which the decisions of two or more actors jointly affect outcomes for resources and the communities that depend on those resources (Ostrom, 2009b). The action arena is the unit of analysis in the institutional analysis and development (IAD) framework, which is used for studying collective action and the effect of rules, attributes of the community and biophysical characteristics on the interaction between multiple actors sharing common pool resources (Kiser and Ostrom, 1982). The interactions between buyers and sellers are connected to a set of adjacent action arenas involving a number of third parties, regulators and stakeholders involved in water planning, the provision and maintenance of infrastructure, conflict resolution, monitoring and enforcement (Kimmich, 2013). Adaptive water institutions are necessary to deliver collective benefits across linked action arenas. Not all transactions are created equal, however. As complexity increases, so does the number of action arenas affected.

Water Transactions: From Simple to Complex

The importance of transaction costs in water reallocation can be grasped by expanding out from relatively simple transactions (between neighboring farms) to more complex transactions across sectors (from irrigation to urban or environmental uses) and jurisdictions (between irrigation districts and states). Pujol et al. (2006) present the conventional perspective on transaction costs as a source of inefficiency in water markets by considering two farms with equal water entitlements but different marginal productivity of water (Figure 2.1). Applying the next bucket of water on one farm will yield more profit than on the other. A shift of an additional unit

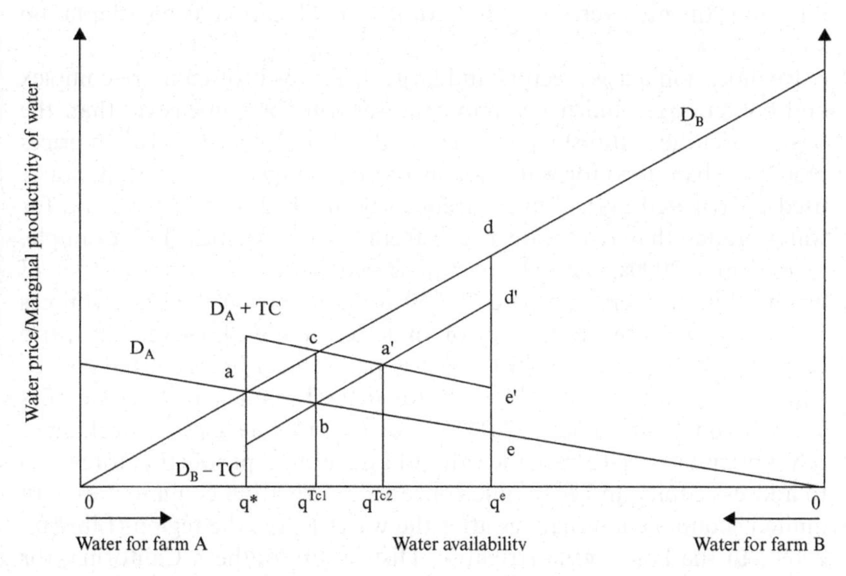

Note: In the initial conditions, water is divided evenly between two farms with different marginal productivities, and the equilibrium level of water reallocation represented by the difference between q_e and q* is constrained by transaction costs faced by both farmers.

Source: Adapted from Pujol et al. (2006).

Figure 2.1 Transaction costs and water reallocation between two farms

of water from one farm (farm A) to the other (farm B) will increase overall productivity with the potential for 'gains from trade': the farm with higher marginal productivity is willing to compensate the other farm such that both are better off. The logic of water trading as an allocation mechanism is to use price signals to coordinate willing buyers and sellers to achieve this outcome. Water rights should be reallocated until marginal productivity is equal across multiple uses. In practice, transaction costs alter the supply and demand curves, raising the costs of supply from the farm with lower productivity (D_A + TC), and decreasing the benefits for the farm with higher productivity (D_B - TC). This suppresses reallocation between the two farms. In other words, transaction costs may reduce, or even eliminate, the scope for beneficial reallocation. But this is not automatically a problem if you shift the focus from efficiency to more comprehensive measures of policy effectiveness. Indeed, Colby (1990b) provocatively posited that transaction costs of water trading in the Western US – viewed as a barrier to reallocation – were still too low to fully address the equity aspects of water allocation. Costly institutional transitions to water rights systems and river basin governance arrangements, coupled with periodic adjustments over time, are critical to reallocation as an adaptation strategy.

Reallocation across sectors and large distances is often more complex and contentious, and hence more transaction costs intensive, than the basic agricultural transfers outlined by Pujol et al. The disparity in prices paid by urban users for water versus those paid by farmers for the water used in irrigated agriculture is frequently invoked as a justification for water trades that reallocate water from farms to cities. For example, in the early 2000s, the City of San Diego paid $225 per acre foot of water while farmers had paid $15.50 for water from the same sources (Libecap, 2005). In the absence of an active market, however, this price differential is misleading. Instead, it is partly a reflection of the cost of delivery, which is comparably high for treated municipal water supplies (Hanemann, 2006; Zetland, 2011). However, cities have proven willing to pay a premium (well above the value of agricultural production foregone) to address equity and economic concerns in irrigation communities with limited economic alternatives after the water leaves the region (Libecap, 2009). In the Palo Verde Irrigation District of Southern California, for example, irrigators paid a flat charge of $52 per acre foot per year in 2005, essentially an administrative charge to recover costs of operating and maintaining irrigation infrastructure. More than 56 per cent of the district land was being used to cultivate alfalfa which is a relatively low value crop. After other production costs are accounted for, profit margins are tight (before subsidies). Against this backdrop of production costs and profits, the Metropolitan Water District of Southern California paid a one-time fee of more than $3000 per acre of land plus $700 per acre foot of water per year to acquire an annual lease to the water. This provides a stark indication of the differing value of water across uses and the premium paid to address the range of community impacts and concerns associated with long-distance transfers from agricultural to urban water use (WGA, 2012). Comparisons between the price paid for agriculture-to-agriculture versus agriculture-to-urban trades vividly illustrate this premium. In the Truckee Basin of Nevada and the South Platte Basin of Colorado, agriculture-to-agriculture trades were $1216 and $4304 per megalitre, respectively, while the corresponding agriculture-to-urban trades were $14 337 and $5285 per megalitre (Grafton et al., 2012, based on Libecap, 2011b, 2011a).

Reallocation across sectors involves potentially higher transaction costs than reallocation within sectors, due to hydrological studies, lawyers' fees and/or administrative procedures, and costly public engagement and conflict resolution efforts as the effects of the trading ripple out beyond the buyer and seller. Groundwater and conjunctive use of surface water and groundwater add layers of complexity and associated information costs to assess the impacts of trading on water use patterns and third parties (Skurray et al., 2013; Blomquist et al., 2004; Ross, 2012). More complex transactions may therefore occur through one-off negotiated exchanges rather than a conventional marketplace, although the Murray–Darling has demonstrated capacity to build an elaborate set of trading rules to facilitate more complex deals, particularly across jurisdictions (although less active trade from agricultural to urban sectors).

Transactions from private economic uses (for example irrigation) to public economic goods (for example environmental flows) entail even more complexity than those between two private economic actors. Part of this is technical. Using water for the environment involves uncertainty and technical challenges. Reallocation to address unmet environmental needs presents novel information gathering and monitoring challenges after irrigation water is returned instream or to wetlands (Garrick et al., 2009). However, a large part of this is political. In the closed or closing river basins of Australia and the Western US, cumulative diversions have exceeded ecological limits for tributaries and downstream deltas; court decisions and legal reform have spurred reallocation from consumptive uses to restore environmental flows. A range of reallocation options have been attempted, including court mandates, administrative decrees or market-oriented voluntary buybacks of entitlements from willing landowners. Voluntary, market-oriented transactions have secured water for the environment after these alternative mechanisms have stalled or failed (Garrick and Aylward, 2012).

Market-based environmental water transactions are therefore even more transaction costs intensive than trading between consumptive uses (for example farms, cities), where transaction costs of environmental water acquisition comprise as much as 70 per cent of total costs (Garrick et al., 2013a; Garrick and Aylward, 2012). Water rights shift from private to public uses, and the environment is a complicated water user with unique timing and volume needs that are assessed using scientific criteria and stakeholder input. Environmental water transactions also involve new organizational forms, such as public and private water trusts, quasi-governmental watershed or catchment organizations, and public agencies at various scales; governments and private organizations act on behalf of the environment in the marketplace, which has required new regulatory conditions, infrastructure and information to ensure the water supplied from previous irrigation users matches environmental water needs at the appropriate times and places.

These reallocation challenges form a spectrum from relatively simple agricultural transactions to complex and often contested transactions from agricultural to urban or environmental uses. River basins encounter multiple trade-offs all at once. This creates pressure to reallocate water among agricultural users and between agriculture and other sectors. Closed river basins face the added challenge of establishing diversion limits and enabling reallocation across irrigation and other consumptive users to restore environmental flows after water has been fully allocated. Intensified competition therefore raises the benefits of reallocation as the scarcity value of water increases, but entails substantial transaction costs due to the widening scope of stakeholders and information involved. The conflict and competition over variable water supplies makes water trading a contentious prospect due to social, economic and ecological interdependencies.

Nature-Related Transactions: Striving for Coherence in Connected Physical and Social Systems

The German agricultural and institutional economist Konrad Hagedorn has coined the term 'nature-related' transactions as those with physical characteristics of natural systems marked by 'jointness and lack of separability, coherence and complexity' (Hagedorn, 2008: 362). In short, natural systems are defined by their ecological and social interactions. Institutions governing nature-related transactions strive for coherence by reducing undesirable environmental and social externalities. This push for 'coherence' underpins efforts to redraw the boundaries of governance arrangements at bioregional scales, such as river basins, to coordinate upstream–downstream interdependencies despite the substantial transaction costs involved with traversing political jurisdictions (see Chapter 5 in this volume). Drawing from his 'institutions of sustainability' framework, Hagedorn proposes two dimensions for classifying nature-related transactions: (1) modularity and decomposability of structures; and (2) functional interdependence of processes. High (low) modularity corresponds with atomistic (complex) transactions, whereas high (low) interdependency entails interconnected (isolated) transactions. Water is multidimensional, with uses that can be both simultaneous (hydropower and recreation) and sequential (upstream and downstream irrigation diversions) with increasing functional interdependency and decreasing modularity as competition intensifies and basins close (Libecap, 2005; Molle et al., 2010). In Hagedorn's terminology, water transactions are 'complex, interdependent' and transaction costs intensive as a consequence.

Water Rights Are Never Complete: Common Pool Resources and Mixed Property Systems

Transaction costs are important because property rights are never complete (Barzel, 1997; Blomquist, 2011). Property rights refer to rules and norms governing interactions between people 'in relation to objects of value' (Challen, 2000; Bromley, 1989). The problem of water is that it supports multiple values. Blomquist (2011) counts at least 16 values attached to water across its multiple uses. In the Great Lakes, 14 beneficial uses have been established by international water quality agreements (Sproule-Jones, 2002). In semi-arid regions, beneficial uses range from farming and fire control to municipal, hydropower and environmental purposes. How to resolve these trade-offs and negotiate the nature-related transactions required for sustainable and adaptive water allocation? Over the past 30 years, Terry Anderson and colleagues at the Property and Environment Research Centre have promoted free market environmentalism as the solution to scarcity and shifting values; free market environmentalism is based on the premise of private, exclusive and secure property rights to align private resource use with the public interest (Anderson and Leal, 2001) – a proposition complicated by issues of equity and sustainability (Raymond, 2003). The neoclassical economic model of property rights is captured by a cost–benefit calculus of institutional change: intensifying scarcity raises the benefits of defining private property rights, while technology (for example, barbed wire for fencing, or remote sensing for irrigation usage) can lower the costs of monitoring and enforcement (Anderson and Hill, 1975). The reality has proven complex and contentious due to the nature of the resource, its multiple uses and values, and the politics associated with defining and adjusting property rights governing it.

Any effort to define private property rights across water's multiple uses and values is destined to be partial and incomplete because of water's common pool resource characteristics: the difficulty of excluding potential users and the need for coordination to generate and sustain water's public goods, which confer indivisible benefits and are plagued by free-riding (Cole, 2002). Exclusion is costly because of the technical and political difficulties of drawing boundaries around water resources and the communities of users with permission to access and withdraw water. Water is not stationary in time or space, which complicates efforts to prevent upstream water users from encroaching on downstream requirements. These exclusion challenges also vary across different economic goods generated by water, some being easier to bound and measure than others, and involve interdependent private benefits (for example, irrigation withdrawals) and collective benefits (for example, environmental flows). If transaction costs were negligible, resource allocation could be continuously renegotiated to achieve optimal outcomes without ever fully specifying property rights. Legal scholar Daniel Cole (2002: 4) notes that 'the existence of a property system necessarily implies the existence of substantial transaction costs'. Instead, water's common pool resource characteristics and multiple uses give rise to a system of 'mixed property rights' (ibid.). Blomquist (2011) concludes that property rights to water must be multidimensional, like the water resources they govern, given the diverse values ascribed to water in these different uses; moreover, decisions about property rights are arrayed across multiple decision-making venues.

In contrast to the private property ideal promoted by market environmentalists, the mixed property rights systems governing water allocation have been conceptualized as polycentric governance arrangements and described as hierarchies or bundles of rights. Different rights and responsibilities are arrayed across multiple levels of social organization, forming what Schlager and Ostrom describe as a bundle of rights – access, withdrawal, decision-making, exclusion and alienation – with different actors holding different sticks (or combinations of sticks) in the bundle (Challen, 2000; Schlager and Ostrom, 1992).

Comparing the three prominent 'panaceas' to water allocation – state, markets and users – Meinzen-Dick (2007, 2014) describes the polycentric arrangements as an 'institutional tripod' in which these three models combine and are mutually reinforcing. The tradable water rights established through market-oriented water allocation reforms often exist alongside state ownership of water and self-governance by water users, such as irrigation districts, to regulate access and use and coordinate provision of necessary infrastructure. In market-oriented cap-and-trade systems, property rights are nested: private tradable rights are conditioned by collective irrigation district rights and state and federal ownership of water and water infrastructure. Changes to water rights at one level have potential knock-on effects at other levels. The unintended consequences and wider impacts of changing use patterns make it unlikely for water markets to be a 'self-maintaining [allocation] system' with a single set-up period followed by a period with 'very little future need for government involvement' (Challen, 2000: 2). This highlights the need to reexamine the relationship between property rights and transaction costs in water allocation in the context of climate change adaptation, environmental flows and the complex multi-dimensional and multi-level trade-offs.

Reinterpreting Coase for Wicked Problems: Property Rights and Transaction Costs

It has been more than 50 years since Nobel laureate Ronald Coase examined the implications of transaction costs for natural resources allocation (Coase, 1960). Coase argued that well-defined property rights address negative impacts of resource use efficiently, regardless of the initial allocation, so long as transaction costs are negligible. He used the example of neighboring landowners – a single farmer and rancher – to illustrate the potential for private bargaining to resolve boundary disputes if property rights are clearly defined. Contemporary resource allocation challenges are comparably much more complex, contested and uncertain, defined as 'wicked problems' for which the framing of the problem is in dispute and the solutions have consequences that are difficult to predict and control (Rittel and Webber, 1973; Batie, 2008; McCann, 2013).

Contrary to the Coase Theorem, the initial allocation of rights has important consequences (Krutilla and Krause, 2010), see Chapter 3 in this volume. The initial allocation establishes dependencies and vested interests, while transaction costs impede adjustments to new information and preferences. Coase knew this and was often misinterpreted, and he urged analysts into the world of 'positive transaction costs' (Coase, 1988). He further warned against presuming that market-based resource allocation is always preferred. Coase and many of his successors recognized the importance of transaction costs, not only those incurred to renegotiate property rights but also those incurred to establish property rights in the first instance (McCann and Easter, 2004). In 1960 the transaction costs world already had multiple layers, with the costs of transferring property rights nested within the costs of creating and maintaining the property rights system.

Transaction Costs Across Multiple Levels of Collective Action, and over Time: From COW to C-WON

Water rights are organized across multiple levels of social organization, and therefore so are the transaction costs of water allocation and institutional change. A multilevel conceptualization of transaction costs and institutional change has been a cross-cutting feature of economic and political economic analysis of public policy and property rights. Both Araral (2013) and Garrick et al. (2013b) synthesize multiple theoretical traditions for analyzing transaction costs and institutional change, by linking insights from **C**oase, Elinor **O**strom and Oliver **W**illiamson; a group of ideas and analytics described by Araral as the 'COW' model. The COW model identifies institutional design attributes that ensure transaction costs are low enough to allow efficient adaptation: property rights must be clear, tradable, perceived as fair, and enforced (Araral, 2013). But how to achieve these conditions in a context of vested interests and institutional inertia? Douglass North is conspicuously absent from the COW model and a historical perspective is only implicit in the model. North highlights the relationship between path dependency and 'adaptive efficiency' which he describes as an 'ongoing condition in which the society continues to modify or create new institutions as problems evolve' (North, 2006). These four theoretical traditions have been applied to water markets and associated allocation policy reforms (Challen, 2000; Marshall et al., 2013; Garrick et al., 2013b). The insights from Williamson, Ostrom and North (WON) form a rounded picture of the dynamic relationship between transaction costs and institutional change. Together, these perspectives set up a framework for understanding the evolution and performance of cap-and-trade water allocation reforms at the intersection of water trading and river basin governance (see Table 2.1). Each is examined in turn below, noting the relevant definitions, determinants and evaluative criteria involved.

Williamson: characteristics of the transaction

For Williamson, transaction costs are friction costs incurred before, during and after the contracting process to collect information, search for willing buyers and sellers, and control opportunistic tendencies of economic actors who face cognitive constraints in maximizing their interests. Contracts are costly to devise, and like property rights, cannot be exhaustively defined across all potential futures, creating potential for

Table 2.1 Transaction costs and the institutional evolution of water markets and allocation policy

Theoretical tradition	Definition of transaction costs	Evaluative criteria	Levels of institutional analysis	Application to water allocation policy
Transaction costs economics (Williamson, 1998)	Information, bargaining and enforcement	Efficiency (transaction costs minimization)	• Embeddedness • Institutional environment • Governance • Resource allocation	McCann and Easter (2004) Easter and McCann (2010)
Path dependency and adaptive efficiency (North, 1990, 1994)	Specifying and enforcing the exchange	Adaptive efficiency (Marshall, 2005)	• Institutional lock-in • Institutional transition • Static transaction	Challen (2000) Heinmiller (2009)
Collective action and common pool resources (Ostrom, 1990)	Costs of collective action (exclusion, monitoring and enforcement)	Robustness, resiliency and sustainability	• Constitutional • Collective choice (transformation) • Operational	Schlager (2005) Schlager and Blomquist (2008)

Notes:

1. Typology of transaction costs in the Williamson tradition has been advanced by McCann et al. (2005). The typology of transaction costs in the North tradition has been developed by Challen (2000) and Marshall (2005, 2013). Marshall (2005) and Carey and Sunding (2001) develop North's criterion of adaptive efficiency; see also Garrick and Aylward (2012).
2. Each tradition emphasizes interactions across levels of action.

opportunism, such as upstream diversions interrupting downstream delivery. Williamson (1998) outlined four levels of social organization: resource allocation (L4), organized by governance mechanisms (L3), which are in turn situated within an institutional environment of laws and rules (L2) and embedded norms (L1). The primary focus of institutional analysis is the 'discriminating alignment' between transactions (L4) and governance mechanisms (L3): the matching of governance mechanisms to the characteristics of the transaction in order to economize on transaction costs.

Water transactions vary in their complexity for several reasons, including the characteristics of nature-related transactions described above. Williamson introduced the concept of asset specificity to characterize transactions in terms of the potential to 'redeploy' the investments in information and technology carried out to conduct a given transaction. Asset specificity is a function of uncertainty, complexity and frequency. Higher asset specificity entails comparably higher transaction costs; nature-related transactions in water allocation are asset specific due to the lack of modularity and high degree of functional interdependence requiring specialized hydrological studies and complex administrative procedures. For example, irrigation districts often established rules to treat water rights as fungible shares but only within the district, which lowered transaction costs for water reallocation within district boundaries and blocked water from leaving the district by requiring a case-by-case review of trading proposals. Thus, a discriminating alignment has formed between the irrigation district and the transactions occurring within its boundaries to minimize transaction costs within its boundaries but not beyond them.

Changes to the governance mechanisms alone may not be sufficient to increase the efficiency of transacting. In the example of the irrigation district, the high asset specificity and relatively low frequency of out-of-district water trades may require institutional change at other levels of social organization to address legal restrictions and underlying cultural perceptions and fears of water leaving irrigation. This multilevel approach provides a nested perspective on water institutions and transaction costs (Easter and McCann, 2010). Environmental water transactions provide another example of high asset specificity prompting new organizational forms and deeper legal and policy reforms. Chapter 4 considers the design challenge of matching organizational forms to complex environmental water transactions in the Columbia Basin where a range of public, private and quasi-governmental entities have entered (and even created) the market to acquire water for the environment. In this context, the emergence of public–private partnerships can be viewed within a transaction costs minimization lens to allow lowest-cost access to relevant hydrological, irrigation, regulatory and economic information. The heterogeneity and associated asset specificity of water transactions, as well as the strongly vested interests in irrigation communities, demonstrate the need for both upfront reforms to the legal environment to streamline water rights and establish diversion limits, coupled with periodic adaptations to navigate the shocks and unforeseen challenges that arise.

North: importance of path dependency

The adaptability of water allocation institutions recalls the observations of Australia's Alfred Deakin about the tenacious grip of vested interests in the Western US irrigation institutions as early as the 1880s, shortly after these institutions formed. The institutions formed to develop water resource infrastructure, and initial allocations created 'dependencies that cannot be maintained' when viewed from the vantage of contemporary efforts to balance production and sustainability in water allocation (Connell, 2007). Transaction costs minimization cannot be divorced from the institutional context and economic history that created the initial assignment of property rights. The impact of history, technology and path dependency is the focus of Douglass North's (1990) analysis of institutional change and economic performance. Institutional change is marked by incremental adjustment punctuated by abrupt transformation. Heinmiller (2009) examined path dependency as a key explanatory variable in efforts to adjust the diversion limits in three overallocated rivers in Australia, Canada and the US; path dependency constrains adjustment due to: (1) vested interests of established water users; (2) sunk costs as water infrastructure becomes obsolete; (3) the contractual nature of property rights; and (4) networks of institutional arrangements at the local, state and interstate levels whose interconnections make changes at one level difficult to isolate from changes at higher or lower levels, producing unintended consequences and cascading impacts (Heinmiller, 2009; Libecap, 2011a).

The Murray–Darling water market reform experience has been viewed through North's institutional economic perspective. Challen (2000) considered the risks of institutional reforms in the mid-1990s to treat water rights as private property. He noted that path dependency would contribute to lock-in by limiting future flexibility to adjust irrigation rights in the public interest. These concerns have been borne out by the reform experience in the Murray–Darling (Crase et al., 2013) and are discussed throughout this book.

Ostrom: transaction costs and institutional change in multiscale commons

Water's common pool resource characteristics render exclusion difficult and costly across multiple levels of action, and over time. Groundwater systems of Southern California were among the first used to illustrate these collective action dilemmas in common pool resources (Ostrom, 1965). Nearly 50 years since Garrett Hardin's tragedy of the commons thesis entered the scholarly and popular lexicon (Hardin, 1968), common pool resource studies have assembled theory and evidence documenting the importance of establishing and maintaining boundaries governing access and use, as well as the conditions favoring the development, monitoring, enforcement and adjustment of such multilevel property rights systems. Transaction costs have a key explanatory role in both the emergence and maintenance of complex property rights systems for common pool resources (CPRs). While institutional design principles associated with successful self-governance of CPRs have received extensive attention, Ostrom lamented the omission of transaction costs from theories and models of collective action (Ostrom, 1990). Like Williamson's transaction costs economics, theories of collective action in the commons are founded on multiple, interacting levels of action (Kiser and Ostrom, 1982). Operational-level actions on the ground are situated within 'progressively more constraining and costly collective choice and constitutional-level actions taken to modify and establish institutional arrangements' (Garrick et al., 2013b: 198). The interaction between operational and collective-choice levels of water allocation is a primary focus of the reform experiences in the Colorado and Columbia Basins, where water rights and river basin governance reforms proceed within prevailing constitutional frameworks for decision-making; by contrast, the Murray–Darling Basin has undergone constitutional-level rule changes governing how decisions are made at the basin level, which has both short and long-run impacts on the transaction costs of river basin governance arrangements aimed at enhancing integration; the constitutional level establishes the rules of decision-making (for example, unanimity, majority, deliberation processes, and so on), which affects the numbers and diversity of actors involved; two key influences on the costs and likelihood of institutional change. Large, more complex and contested water allocation challenges are encumbered by high coordination costs due to the number of actors involved and the different decision rules (for example, consensus versus majority) affecting the costs of securing an agreement (Ostrom et al., 1999).

Transaction costs and the calculus of institutional change

Institutional change occurs when the benefits of a proposed change exceed the costs incurred for a 'winning coalition' of stakeholders or political interests capable of adopting and implementing a rule change. This is a daunting challenge when the costs are upfront, concentrated and well defined, while the benefits are delayed, distributed and uncertain

(Harrison, 2013). The political tensions over water buybacks in the Murray–Darling Basin are a case in point, where efforts to acquire water entitlements for the environment have stalled despite bipartisan political support due to well-organized resistance by those with irrigation interests who bear a disproportionate share of the costs of change (Crase et al., 2013).

As noted above, the efficiency school of property rights economics viewed the evolution from open access to private property as a natural progression to maximize economic welfare (Anderson and Hill, 1975; but see Rose, 1990; Schorr, 2012). Upfront costs that are highly concentrated on the potential losers from a reform may thwart (otherwise) beneficial reallocation and associated rule changes before they get off the ground. Upfront costs are associated with the heterogeneity of the group, information base and norms of reciprocity (Ostrom, 1990). Large, contested river basins involve diverse stakeholders, complex social and ecological interactions, and complex cross-scale institutional linkages that favor the status quo.

The calculus of institutional change is formalized by Basurto and Ostrom (2009: 44–45). They consider the costs and benefits of a rule change for both individual users and groups. Each individual (*i*) faces a decision to whether to support a rule change based on the calculus:

$$D_i > (C1_i + C2_i + C3_i)$$

where:
D_i = benefits of rule change for the individual
$C1_i$ = upfront costs of developing and adopting new rules
$C2_i$ = short-term costs of implementing new rules
$C3_i$ = long-term costs of monitoring and maintaining a self-governed system over time.

If the benefits of a rule change outweigh the costs for at least one user, then the question becomes whether there is a 'minimum coalition' of '*k*' users, for which:

$$D_k > (C1_k + C2_k + C3_k)$$

The size of the minimum coalition is a function of the collective-choice rules, ranging from unanimity to individual fiat. The collective-choice rules will affect the structure of the costs. For unanimous agreement or supermajorities, the upfront costs (*C1*) will be high, but the 'winning coalition' will be broad and more likely to observe the rules and decrease the short-term (*C2*) and long-term (*C3*) costs of implementation, monitoring and maintenance. While a 'winning coalition' comprising elites may economize on upfront costs, the perceptions of unfairness may limit compliance and raise the subsequent costs. These trade-offs between types of transaction costs have 'intertemporal' dimensions: decisions about policy design and property rights reform that influence the stream of future costs, the subject of the next section and Chapter 3.

Transaction costs of scaling up

Common pool resource studies also highlight issues of scaling up beyond user self-governance to address trade-offs in larger, nested systems. A multiscale resource such as water also involves tensions between different communities of interest: user groups, sub-basins, states, national governments or international actors. Schlager and Blomquist (2008) explain the persistence and proliferation of polycentric river basin governance arrangements rather than integrated basin-wide trade-offs due to bounded rationality and transaction costs. Bounded rationality refers to cognitive limitations, while transaction costs reinforce these limitations by making decisions and institutional development costly. Together, bounded rationality and transaction costs favor satisficing decisions to identify acceptable, often sub-basin, institutional arrangements because the benefits of comprehensive basin-wide solutions may not offset institutional development costs for communities of interest whose needs can be met without decision-making at the whole-of-basin level. These transaction costs of bioregional 'coherence' are the focus of multiple streams of institutional analysis of polycentric governance arrangements, including 'local public economies', institutional collective action and federalism (see Chapter 5).

TRANSACTION COSTS TYPOLOGIES AND MEASUREMENT

Recognition of the importance of transaction costs has been accompanied by efforts to measure and account for them in analysis of market-based water reallocation and associated institutional reforms (Brown et al., 1992; MacDonnell, 1990; Colby, 1990b, 1990a). This parallels (and lags) efforts to understand transaction costs and growth in the wider economy. An assessment by Wallis and North (1986) estimates the changing size of the transaction costs sector as a proportion of the US economy from 1870 (24 per cent) to 1970 (56 per cent), an expansion driven partially by the transition from manufacturing industries (production costs intensive) to service industries (transaction costs intensive) of an increasingly information-based economy. The study represented one point of departure for empirical measurement of transaction costs and also sparked controversy over the underlying conceptual definitions and boundaries guiding measurement.

In an important article on transaction costs and environmental policy, McCann et al. (2005) invoke management expert Peter Drucker's adage: what gets measured gets managed. Their analysis identifies conceptual and methodological issues in transaction costs accounting by defining transaction costs and considering the typology, chronology and distribution of transaction costs across the policy and implementation cycle. This section outlines measurement challenges in a transaction costs world by defining transaction costs, the main types of transaction costs and their interactions, the relationship between transaction costs and technology, and commonly used methodology and metrics.

Defining Transaction Costs So They Can Be Measured

Measuring transaction costs requires a clear definition of what counts as transaction costs (McCann et al., 2005). The focus and boundaries of the analysis will dictate the types of transaction costs measured and analyzed in a given study (ibid.). Therefore the starting point is a broad and inclusive definition capable of investigating complex nature-related transactions, mixed property rights systems and issues of institutional change. Although definitions have proliferated, none encompass the full range of issues. Instead, three existing definitions together provide a rounded perspective on transaction costs. For the purposes of conceptual clarity and policy analysis, transaction costs refer to resources (financial, staff, time, and so on) required to:

1. address collective action dilemmas of exclusion and coordination in the governance of common pool water resources (Cole, 2002);
2. define and manage property rights, including resources to 'define, establish, maintain and transfer property rights' (McCann et al., 2005: 530) to access, withdraw, exclude, manage and alienate water (Schlager and Ostrom, 1992); and
3. enable institutional development and adaptation for complex, multiscale problems, including resources to 'define, establish, maintain, use and change institutions and organizations and define the problems that these institutions and organizations are intended to solve' (Marshall, 2013: 188).

An Analytical Framework: Types and Interactions

In light of the nature-related transactions and mixed property rights systems governing water, water markets and associated river basin governance reforms cannot be viewed as 'self-maintaining' according to Challen (2000: 2). Institutional change is needed to enhance adaptation and build adaptive efficiency, a criterion broadly defined as the capacity to solve problems over time, and comprising:

1. 'Sequentially more complex institutional innovations (Carey and Sunding, 2001: 291) to lower transaction costs.'
2. Incentives for decentralized institutional capacity-building to explore alternative problem solving strategies and cover associated transaction costs (when lowering them is difficult) (Marshall, 2005; North, 1990).

The broad definition of transaction costs above is geared toward understanding institutional change across the full chronology of institutional development, implementation and adaptation, which corresponds to three interacting categories of transaction costs, as elaborated by Ray Challen and Graham Marshall (Challen, 2000; Marshall, 2005, 2013):

- static transaction costs;[3]
- institutional transition costs;[4] and
- institutional lock-in costs.[5]

Static transaction costs (Figure 2.2) refer to the costs of transacting within the prevailing institutional setting (Challen, 2000). The transaction costs of water reallocation within existing institutional arrangements might include the search for willing buyers and sellers, negotiation of contracts and pursuit of administrative approvals, followed by monitoring and enforcement of the contract. However, institutional change is necessary to set up water markets and adjust them periodically. The notion of adaptive efficiency proposed by Carey and Sunding (2001) suggests that static transaction costs are expected to decrease through ongoing institutional innovations, all else being equal. However, several factors (unintended consequences, irreducible uncertainty, shifting preferences and technologies, and altered demand and supply conditions) may cause static transaction costs to stabilize or increase over time, particularly in the absence of strategic reinvestment in institutional transition costs.

Institutional change involves two additional categories of transaction costs: institutional transition costs and institutional lock-in costs. Transition costs are the costs of changing from the status quo to a new

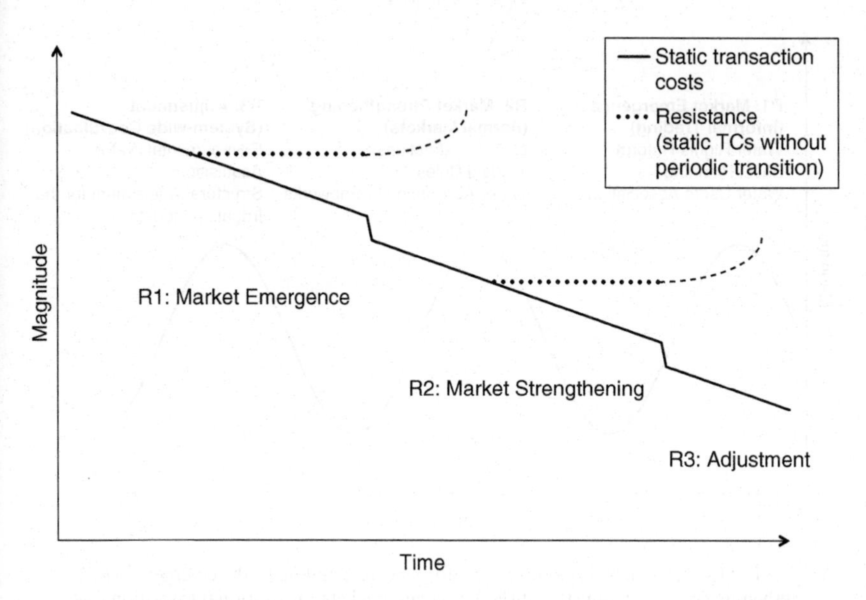

Note: Static transaction costs refer to the costs of transacting within the prevailing institutional setting. The average static transaction costs are expected to decline due to efforts to clarify property rights, build information systems and improve technologies and governance arrangements for monitoring, enforcement and conflict resolution. The decreasing trend may encounter resistance or temporary reversals in direction when conflicts, new preferences, and so on develop, requiring sequential investments in additional institutional transitions.

Figure 2.2 Static transaction costs

institutional setting. Institutional transitions for market-based water allocation reforms include the establishment or revision of sustainable diversion limits, creation or adjustments to water rights systems, water users associations and regulations governing water trade (Figure 2.3). Institutional transitions have their analogue in Williamson's third level of social analysis – the legal environment – and are the realm of policy design. Finally, institutional choices that limit flexibility to respond to new information, shifting preferences or technological change increase the costs of future institutional transitions (Challen, 2000). Intertemporal costs have been described as institutional lock-in costs by Marshall (2005) because they require reversing historic decisions (Livingston, 2005). The privatization of water rights is one potential example because it raises the future costs of institutional transitions that would return some of those rights into the public domain.

Challen (2000) and Marshall (2005, 2013) note that static transaction

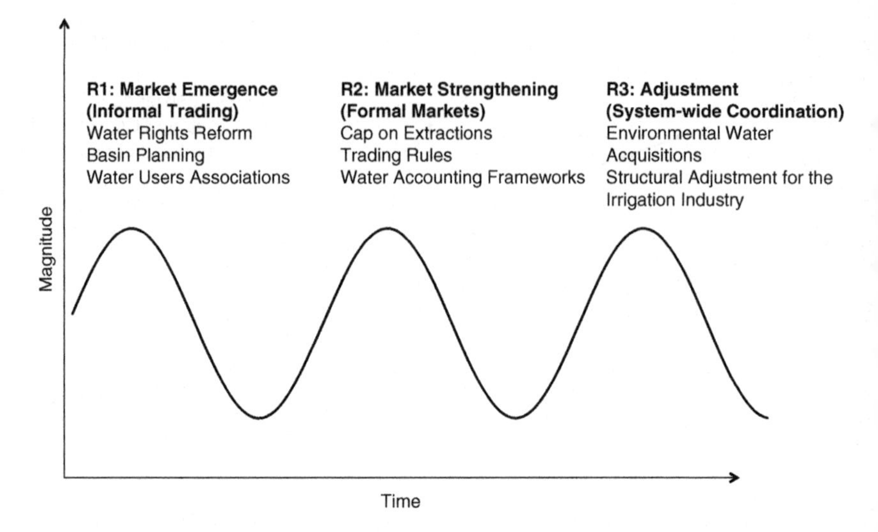

Note: The figure depicts three stages of reform corresponding to the emergence (R1), strengthening (R2), and adjustment (R3) of water markets. Institutional transition costs refer to the costs associated with shifting from the institutional status quo to a new arrangement. Their magnitude is expected to fluctuate cyclically in response to pulses of investment as policy windows open for water allocation reform.

Figure 2.3 Institutional transition costs

costs are the most commonly measured, in part because the boundaries around them are more concrete. For example, it is possible to estimate or survey the costs incurred for hydrological studies and legal fees associated with administrative procedures (for example Colby, 1990b). The transition costs of market-enabling legal reforms have received some attention, but precise quantification is rare and potentially misleading (McCann and Easter, 2004). Garrick et al. (2013b) argue that institutional transitions should not be considered as a single set-up period with a start and end point because water markets are not self-maintaining. Instead, institutional transition costs can be considered as a sequence of periodic transitions with investments in water market reforms followed by periods of implementation and policy learning before new obstacles or opportunities for gains from trade drive future transitions.

Institutional lock-in costs have been considered in conceptual or qualitative discussions of path dependency (Libecap, 2009; Harris, 2011), although Libecap has analyzed the Owens Valley transactions in detail. Institutional transition costs are expected to follow a policy cycle and fluctuate over time during set-up (water rights reform for informal water trading), consolidation (strengthening enabling conditions in formal water markets) and adaptation (for example, adjusting diversion limits). Institutional lock-in costs link institutional transition costs across each period, wherein short-run decisions in the initial phases of reforms to economize on institutional transition costs (for example, grandfathering historic water users) may diminish future institutional flexibility and thereby significantly increase the costs of future institutional transitions. Pulses of investment in institutional transition costs are expected to be triggered, in part, by on-the-ground barriers that impede otherwise beneficial water trades, coupled with policy windows created by crises.

Typologies of transaction costs represent an early step toward a functional understanding of transaction costs in relation to environmental, social and economic characteristics. These measurement challenges reflect more fundamental theoretical and analytical weaknesses to link transaction costs with political economic factors, technology and transformation costs. In the case of the former (political economy), transaction costs minimization may not be the overriding policy objective for many stakeholders; instead rent-seeking and maintenance of vested interests dominate, as when an irrigation community or environmental water users organize to protect or increase their access to water. In the case of the latter (technology and transformation costs), the boundaries between transaction costs and transformation costs are thick and overlapping, requiring the analyst to distinguish between them in the context of resource allocation (Krutilla and Krause, 2010). The typology of transaction costs suggests that alternative market-based water-policy design options involve trade-offs and interactions among types (and magnitudes) of transaction costs incurred, the distribution of transaction costs among multiple parties, and the incidence of transaction costs across multiple phases of reform (Krutilla and Krause, 2010; McCann et al., 2005). For example, economizing on the institutional transition costs associated with setting up the market-enabling policy framework (for example cap-and-trade) may raise the static transaction costs if rights or trading rules are unclear. Likewise, institutional lock-in costs may increase if tradable water rights are designed in a way that reduces flexibility to adjust water rights in response to changing environmental preferences or long-term variability.

Technological Change and Production Costs

North (1994) argued that transaction and transformation costs interact to shape the trajectory of economic performance over time and space. The distinction and interaction between transformation and transaction costs merits additional scrutiny. The literature on transaction costs and environmental policy analysis has consistently emphasized the importance of considering transaction costs alongside the full range of costs and benefits, including transformation costs, or production and abatement costs (McCann, 2013; McCann et al., 2005; Marshall, 2005, 2013). Krutilla and Krause (2010: 272) note however that the boundary between production and transaction costs is indistinct and depends on the nature of the transaction: 'transaction costs and production costs can range from mutually exclusive to overlapping'. Studies of water markets have been critiqued for conflating transaction and transformation costs in two key ways. First, there are the costs of infrastructure needed to store water and to transport water from seller to buyer. Second, there are the costs of water buybacks and irrigation efficiency schemes designed to reduce cumulative diversions in overallocated rivers and regions. For example, Hearne and Easter (1997: 188) describe investments in water storage and distribution systems as a form of transaction costs associated with the 'physical infrastructure needed to measure and transport water'. The second set of costs typically represent a public or private expenditure to reduce the consumptive use by irrigators. Both would traditionally be considered a form of transformation costs.

Regardless of the assignment between transaction and transformation costs, Krutilla and Krause (2010) urge the analyst to treat them as mutually exclusive, which can be accomplished by considering the goods and services in question (Garrick et al., 2013b). While it is important to consider the full range of costs and benefits – and their distribution – in evaluating the evolution and design of water allocation institutions, some forms of transformation costs associated with water allocation are constant across alternative policy and institutional options. It is important to pinpoint which transformation costs vary across alternatives, such as the decision to restore environmental flows through direct acquisition of water rights versus the conservation of water via irrigation efficiency improvements (with the latter incurring higher transformation costs). In previous research, colleagues and I have noted that the contrast 'of infrastructure-driven (transformation-costs intensive) versus market-based (transaction-costs intensive) policy options illustrates how transformation and transaction costs interact with the political economy of water allocation to influence the outcome of reforms' (Garrick et al., 2013b: 197). In other words, it is often more politically acceptable to adopt supply-side and/or infrastructure-intensive water allocation options for which the benefits are concentrated and the costs widely distributed, than the opposite, for example market-based buybacks which distribute benefits broadly and concentrate costs on the irrigation sector. Here my goal is to make transaction costs more visible and explicit in the analysis of institutional change and to explore the path dependent interactions of technology and institutions in water allocation reform. A full treatment of transformation costs and cost-effectiveness analysis is outside the scope of my argument.

Methodology and Metrics for Transaction Costs Analysis

McCann et al. (2005) establish a framework and typology for transaction costs measurement that has guided a range of empirical measurement studies over the past decade. Measurement relies on boundaries to delimit what counts as transaction costs through all stages of policy design and implementation and across all actors. Empirical typologies include the type of transaction costs, market actors and timing to capture the *ex ante* and *ex post* costs through the full policy cycle. The typology includes multiple categories – policy enactment and implementation – as well as subtypes within each category (Table 2.2). In the context of market-based environmental water allocation, policy enactment costs include enabling reforms of water rights institutions to define, transfer and manage water rights in a cap-and-trade system that allows transactions for environmental flow restoration purposes. Enactment costs also entail the development of nested institutions to organize public and private efforts for planning, water banking, and administrative capacity at multiple levels. The subtypes of implementation costs include project-level costs of transaction prioritization and planning, water rights due diligence, negotiation and price discovery, administrative fees and processing, conflict resolution, monitoring and enforcement, and financing (Garrick and Aylward, 2012).

Empirical measurement and modelling of transaction costs and institutional change have become a common part of water market evaluation (for example, Garrick and Aylward, 2012; Colby, 1990b, 1990a). Three approaches are common: surveys of transaction participants, including stated and revealed costs; audits of financial (or other) expenditures; and mathematical models or experiments to identify threshold-level expenditures beyond which the gains from trade are negated. Metrics have accounted for transaction costs in terms of time and staffing levels, expenditures and opportunity costs, administrative processing time, and the number of people and/or meetings involved in collective choice and operational-level actions. However, the expansive definitions of transaction costs have given rise to a growing recognition that some types of transaction costs either cannot or should not be measured.

Table 2.2 Transaction costs typologies: categories in collective action, environmental policy analysis and water markets

Collective action	Environmental policy	Water markets
Institutional transition costs	Research and information	River basin development, planning and closure (cap) Hydrologic and socio-economic studies
	Enactment or litigation	Water rights reform (adjudication, conflict resolution, rules) Establishment or reform of water user associations
	Design and implementation	Modification to storage and distribution Licensing systems Trading rules and registries Price discovery (auctions, tenders, brokerages) Water accounting system
Static transaction costs	Support and administration	Transaction planning Identification of buyers/sellers Administrative review (for example, injury analysis)
	Contracting	Water rights due diligence
	Monitoring and detection	Water use accounting
	Prosecution and enforcement	Compliance monitoring and enforcement Dispute resolution
Institutional transition costs	Adaptation or replacement	Revise cap; adapt water rights and water user association rules; acquire water rights for the environment if cap is revised downward
Source: Adapted from Marshall (2005, 2013)	Source: Adapted from McCann et al. (2005), Marshall (2013)	Source: Adapted from Garrick and Aylward (2012), McCann and Easter (2004)

Note: Theories of collective action have identified different types of transaction costs in the environmental policy and water market context. These typologies identify the distinction and interaction of institutional transition costs (for example collective choice, decision-making) and static transaction costs (for example implementation, monitoring and enforcement).

Source: Garrick et al. 2013b.

A comprehensive cost effectiveness study that accounts for lock-in, transition and static transaction and transformation costs is useful as a heuristic device to aid decision-making about alternative institutional options even when some elements cannot be quantified (Marshall, 2013).

EVIDENCE

Evidence on transaction costs and water allocation reform remains limited, despite growing efforts to measure different types of transaction costs for different groups of actors (buyers, sellers, regulatory bodies, third parties, and so on). Empirical studies can be summarized by some broad trends. First, the magnitude of transaction costs is typically low for completed transactions ranging from 6 per cent to 29 per cent of total costs for consumptive use transactions. These results are potentially misleading, revealing a conundrum for measuring the (prohibitive) transaction costs when a transaction cannot be completed due the transaction costs exceeding the benefits. Second, the frameworks and methods for transaction costs measurement and analysis have become more sophisticated to examine more complex types of water transactions (for example environmental water acquisitions) and integrate transaction costs parameters in mathematical models of water trading potential. Third, the maturing analytical frameworks are being used to identify determinants of transaction costs, including the institutional, technological, and biophysical factors influencing the magnitude and incidence of transaction costs, and the relationship between transaction costs and institutional change. Finally, transaction costs measurement has benefited from conceptual advances and theoretical insights gained from diverse geographic contexts where water markets have been attempted or studied, and from transaction costs analysis in other fields of environmental governance and natural resource allocation. These trends are unpacked below.

The resulting evidence on water markets and transaction costs suggests that static transaction costs are a function of incomplete property rights and third party impacts, imperfect information on hydrology and legal characteristics, and the highly contested political economy of water reallocation. Search, bargaining and enforcement costs are likely to prove a barrier to trading in incipient water trading programs (Bauer, 1997; Challen, 2000; Colby, 1990b, 1990a). In the Murray–Darling water market these factors contributed to price dispersion in early trading activity, that is, prices varied for water rights despite similar characteristics of the water rights (Challen, 2000). In the Western US, the high risk of protests and administrative delay has suppressed trading activity because costs were expected to be prohibitively high or to lead to rejected applications (Lund, 1993; Baker and McKee, 2000).

High fixed costs of water trading have created economies of scale which bias reallocation activity toward transactions involving relatively large volume and/or transactions within existing networks and jurisdictions, as exemplified by the Westlands Water District of California (Carey et al., 2002; Ruml, 2005; McCann, 2009). As a consequence, informal spot market trades have been prevalent due to jurisdictional barriers and infrastructure constraints on cross-scale trading activity (Thompson, 1993; Bauer, 1997; Carey et al., 2002; Ruml, 2005). Although water allocation authority is vested at the state level in the US and Australia, transaction costs vary as much within states as across them due to the importance of local time- and place-specific information generated through field-based administration and adjudication of water rights (MacDonnell, 1990; Garrick and Aylward, 2012). More complex water transactions, such as environmental water reallocation, have entailed comparably high transaction costs as a percentage of total costs, particularly to reallocate water up the institutional hierarchy from individual users into the public domain (Challen, 2000; Garrick and Aylward, 2012).

Three additional observations emerge from a review of the evidence. First, quantitative analysis of transaction costs has focused on the static transaction costs during the emergence of water trading activity, but longitudinal assessments are rare even when barriers appear significant and trading activity remains limited. Early studies compared the costs of transactions across and within local and state-level water markets in the Western US, Australia and Chile. In regions without trading activity, experimental and mathematical modelling studies assessed transaction costs to identify the gains from trade needed to overcome different threshold levels of fixed and variable transaction costs (Pujol et al., 2006; Garrido, 1998). Despite these efforts, simulations of water markets regularly ignore transaction costs or merely acknowledge them as exogenous (but see Erfani et al., 2014). As noted above, less attention has been given to the institutional transition costs of setting up the market-enabling institutional framework (but see McCann and Easter, 2004) or the institutional lock-in costs associated with path dependency. The latter have primarily been addressed through historical case studies of individual projects (for example Owens Valley, California; Libecap, 2009) or by considering the impacts of past transactions on contemporary and future water reallocation via mathematical modeling (Challen, 2000) or qualitiative comparisons (Libecap, 2009; Garrick and Aylward, 2012; Heinmiller, 2009; Challen, 2000).

Transaction costs vary across and within jurisdictions in response to the complexity, conflict and externalities of the transaction projects, and the local and state institutional capacity to manage such challenges (MacDonnell, 1990; Colby, 1990b, 1990a; Archibald and Renwick, 1998). Colby (1990b) surveyed transfer applicants and legal professionals to identify average costs and cost as a proportion of total costs (inclusive of the 'policy-induced' transfer approvals required for statutory changes in use). Transaction costs were $187, $54 and $66 per acre foot on average in Colorado, New Mexico and Utah, respectively (for $91 per acre foot and 6 per cent of total costs across the multistate study area) while administrative processing times ranged from four to 29 months. Brown et al. (1992) tracked applicant costs for 201 transfers with transaction costs varying from $0.06 to $1100 per acre foot (average of $135 per acre foot); that study built on prior work by Nunn (1990) who surveyed applicants for 87 transactions to track attorney fees, court cases, hydrological expenses and filing and publication fees. Other studies focused on administrative costs, including Howitt (1994) in California (8 per cent of total costs); and Hearne and Easter (1995, 1997) in Chile, finding $0.07/m^3$ in the most active trading area of the Limari Valley to $0.23/m^3$ in the less active Elqui Valley. This review demonstrates local variation (within countries) in water market performance (see Table 2.3).[6]

Transaction costs research in the Australian market has been limited until recently. Challen (2000) applied an econometric model of the effect of imperfect information on water prices, estimating total transaction costs in the South Australian Murray–Darling water market of AU$12–125/ML (megalitres) – of which AU$60–$70/ML is out-of-pocket costs and the rest is the implicit cost of information – or 3–29 per cent of the average transaction price (Challen, 2000). Crase et al. (2000) model buyer and seller behavior to capture the effect of 'policy flexibility' associated with weakened property rights to water, indicating a willingness to pay higher prices for more secure property rights. The Allen Consulting Group (2006) provides the first empirical analysis of transaction costs in the water market that is based on both public and private expenditures. Nevertheless, its scope is limited to a subset of private transaction costs and government transaction costs that are meant to be an 'indicative. . .rather than a definitive summary'. Private transaction costs ranged from 3 to 21 per cent of total costs for an average-sized transaction; government transaction costs – the costs of managing, but not establishing, the water trading system – totalled more than AU$11 million each year in New South Wales (NSW) and Queensland. Empirical studies are limited to snapshots, typically early in the emergence and development water markets. Studies concentrate

Table 2.3 Transaction costs in emerging water markets: static transaction costs

	Author(s)	Year	Country	Economic good	Transaction costs (% of total costs)	Transaction costs (nominal USD/m³)		
					Average or range	Low	Average	High
1	Colby	1990	US	Quantity	6	$0.05[a]	$0.07	$0.15
2	Brown et al.	1992	US	Quantity	~13	<$0.01[b]	$0.23	$1.12
3	Howitt	1994	US	Quantity	8			
4	Hearne and Easter	1995	Chile	Quantity	7–23	$0.07[c]		$0.23
5	Garrido[d]	1998	Spain	Quantity	8–12		$0.024	
6	Challen	2000	AUS	Quantity	3–29	$0.01		$0.13
7	Garrick and Aylward	2012	US	Environmental flow	12–70	<$0.01[e]	$0.01	$0.03

Notes:
Results are based on McCann and Easter (2004) and were supplemented through direct consultation and analysis of each study's results. Water volume units for Brown et al. and Colby studies were converted from acre feet to cubic metres, while Challen study units were converted from megalitres to cubic metres.

a. 'Range' refers to minimum and maximum average transaction costs per state jurisdiction for Colorado, New Mexico and Utah. 'Average' refers to the entire study area.
b. 'Range' refers to minimum and maximum average transaction costs per sub-basin from a sample of eight. 'Average' refers to the entire study area.
c. 'Range' refers to minimum and maximum average transaction costs per region.
d. Refers to a threshold level of transaction costs beyond which gains from trade are negated.
e. 'Range' refers to minimum and maximum average transaction costs per sub-basin from a sample of 13. 'Average' refers to the entire study area. Water volume units were cubic feet per second, which were converted to a volume (m^3) by aggregating the flow rate over a traditional 180-day irrigation season.

principally on transactions involving consumptive uses (for example agriculture-to-agriculture).

Second, empirical analysis has addressed increasingly complex water allocation challenges and economic goods from water quantity (Hearne and Easter, 1997; Colby, 1990b, 1990a) to water quality (McCann and Easter, 1999) and, most recently, environmental flows (Garrick and Aylward, 2012). Conceptual typologies have supported increasingly sophisticated methodological innovations and metrics in transaction costs accounting (McCann et al., 2005; Marshall, 2013). Initial studies relied on surveys and audits of expenditures focused on administrative reviews and fees, legal expenditures for conflicts, and hydrologic analysis. More recent studies have relied on non-market valuation techniques (McCann et al., 2005; Peterson et al., 2011; Speelman et al., 2010) and theoretical and empirical mathematical models to identify thresholds levels beyond which transaction costs erode gains from existing or potential trade activity (Garrido, 1998; Pujol et al., 2006).

Finally, scholars have examined the determinants and consequences of transaction costs in market-based water allocation policy. Most studies have applied the Williamson (McCann and Easter, 2004) and North approaches (Challen, 2000; Heinmiller, 2009) to institutional analysis, followed more recently by Ostrom-based analysis of water rights (Schlager, 2005) and integrated river basin management (Schlager and Blomquist, 2008). The evidence indicates that water's physical characteristics, incomplete property rights, information costs and highly contested political economy drive up search, bargaining and monitoring costs (Bauer, 1997; Bjornlund, 2004; Challen, 2000; Libecap, 2009; Colby, 1990b, 1990a) and contribute to the risk that trading activity will be suppressed (Baker and McKee, 2000; Lund, 1993; Young, 1986).

ADAPTIVE EFFICIENCY: DEFINING SUCCESS IN A TRANSACTION COSTS WORLD

Water allocation reform in closed river basins is a challenge of adaptation and reallocation. Transaction costs analysis provides a lens to understand stability and change in water allocation in a context of path dependency, nature-related transactions and complex, contested economic goods with interdependent public and private values. The relationship between transaction costs and water rights reform was initially viewed through the lens of neoclassical efficiency: well-defined, private and tradable rights that maximize economic welfare when scarcity elevates the benefits and technological innovation reduces the transaction costs of property rights reform (Anderson and Hill, 1975). This logic of institutional change has been interpreted and applied to the water sector by Saleth and Dinar (2004) in their contention that water sector reforms (including water allocation mechanisms) would be adopted when the opportunity costs of misallocation (that is, foregone benefits) outweigh the transaction costs. Economic efficiency is the criterion for success, and efficiency is defined as the 'least-cost path' to reach a fixed policy target. This evaluative criterion takes the institutional structure as constant (see Bromley, 1982 for a careful critique); dynamic efficiency captures adjustments to supply and demand, but also within the prevailing institutional setting.

Adaptive efficiency treats institutional design as an independent variable, rather than a constant, and one that is integral to the analysis of economic performance over time. Complex property institutions have evolved in response to the mixture of public and private values tied to water. These property institutions need to adjust periodically, that is, to reallocate water rights as values shift and to alter the river basin governance arrangements used to establish cumulative diversion levels at nested hydrological scales. In contrast with neoclassical economic measures of efficiency, adaptive efficiency refers to efficiency over the long term. Adaptive efficiency is an 'ongoing condition in which the society continues to modify or create new institutions as problems evolve' (North, 2006: 169) in the context of complexity, uncertainty, shocks and pervasive feedbacks (see also North, 1990, 1994; Marshall, 2005; Marshall et al., 2013). Considering the political economy and path dependency of water allocation, Carey and Sunding describe adaptive efficiency as 'sequentially more complex institutional innovations to lower transaction costs' (Carey and Sunding, 2001: 291).

Selecting adaptive efficiency as an evaluative criterion implies a number of corollaries for institutional analysis. First, there is the need to focus on institutional change in the face of uncertainty. Second, there is the need to account for informal norms, which form over long periods, and emphasize the role of path dependency, see Chapter 3 in this volume. North cautions that there is no short cut to establish these enabling institutional conditions, and therefore the development of water markets and river basin governance arrangements require sustained and sequenced institutional transitions. Third, the trials and experimentation involved are likely to require institutional diversity and polycentric governance arrangements rather than pure ideal types of markets, states or users (Ostrom, 2009b). Such polycentric governance arrangements are unlikely to strive for efficiency (in terms of transaction costs minimization) as the sole policy objective. In this context, the calculus of institutional change elaborated by Demsetz and by Anderson and Hill needs to be modified and extended for complex nature-related transaction costs, mixed property systems and ongoing institutional adaptations involving: institutional transition costs up front and over time and the associated interplay of transaction costs and incentives, for both individuals and collective choice bodies. This concept of transaction costs and adaptive efficiency reveals the importance of collective choice institutions and coalitions required to adopt and implement rule changes, which aggregate the individual preferences through collective choice rules governing who gets to decide and how (Basurto and Ostrom, 2009).

The focus on adaptive efficiency as an evaluative criterion highlights the sequential process of market-based water policy reform and the need for sustained investment in institutional transition costs through multilevel collective action among users and institutions across and within jurisdictional boundaries (see the Columbia Basin, Chapter 4 in this volume; Garrick and Aylward, 2012). In sum, the 'state of the art is always provisional' (Blackbourne, 2006). Above all, adaptive efficiency is predicated on the notion of market-oriented water allocation reform as a phased process comprising incremental change with periodic transformation, which starts from a recognition of path dependency, lock-in and their dynamics – the subject of the next chapter.

NOTES

1. Including both transaction costs and transformation costs, as described below.
2. Palo Verde Irrigation District, n.d. History. Available at: http://www.pvid.org/history.html. Accessed October 2012.
3. Static transaction costs are broadly equivalent to *C2* and *C3*, see above (Basurto and Ostrom, 2009).
4. Institutional transition costs are broadly equivalent to *C1*, see above (Basurto and Ostrom, 2009).
5. Adapted from Challen (2000) by Marshall (2005, 2013) as part of a comprehensive cost-effectiveness framework. Note that institutional lock-in costs lack a clear analogue with the three types identified by Basurto and Ostrom (2009), but see chapter 3 for a proposed extension to this calculus of institutional change to capture the intertemporal costs imposed by past decisions.
6. Even within a basin, or a region served by inter-basin transfers, there is tremendous variability as illustrated by the Colorado River. The Colorado-Big Thompson Project (served by an inter-basin transfer from the Colorado River) is one of the most active water markets in the Western US, while other states and districts in the Colorado Basin have minimal trading or confront high transaction costs despite intense competition for water. See Howe and Goemans (2003).

REFERENCES

ALLEN CONSULTING GROUP. 2006. Transaction costs of water markets and environmental policy instruments. Melbourne.

ANDERSON, T.L. and HILL, P.J. 1975. The evolution of property rights: a study of the American West. *Journal of Law and Economics*, 18, 163–79.

ANDERSON, T.L. and LEAL, D.R. 2001. *Free Market Environmentalism*. New York: Palgrave Macmillan.

ARARAL, E. 2013. A transaction cost approach to climate adaptation: insights from Coase, Ostrom and Williamson and evidence from the 400-year old *zangjeras*. *Environmental Science and Policy*, 25, 147–56.

ARCHIBALD, S.O. and RENWICK, M.E. 1998. Expected transaction costs and incentives for water market development. In: EASTER, K.W., ROSEGRANT, M. and DINAR, A. (eds), *Markets for Water: Potential and Performance*. New York: Kluwer Academic.

BAKER, K. and MCKEE, M. 2000. Increasingly contested property rights and trading in environmental amenities. *Land Economics*, 76, 333–44.

BARZEL, Y. 1997. *Economic Analysis of Property Rights*. Cambridge: Cambridge University Press.

BASURTO, X. and OSTROM, E. 2009. Beyond the Tragedy of the Commons. *Economia delle fonti di energia e dell'ambiente*, 52, 35−60

BATIE, S.S. 2008. Wicked problems and applied economics. *American Journal of Agricultural Economics*, 90, 1176–91.

BAUER, C.J. 1997. Bringing water markets down to earth: the political economy of water rights in Chile, 1976–1995. *World Development*, 25, 639–56.

BJORNLUND, H. 2004. *What Impedes Water Markets*. Sydney: Moin & Associates.

BLACKBOURNE, D. 2006. *The Conquest of Nature: Water, Landscape and the Making of Modern Germany*. London: Norton.

BLOMQUIST, W. 2011. A political analysis of property rights. In: COLE, D.H. and OSTROM, E. (eds), *Property in Land and Other Resources*. Cambridge, MA: Lincoln Institute of Land Policy.

BLOMQUIST, W., SCHLAGER, E. and HEIKKILA, T. 2004. *Common Waters, Diverging Streams: Linking Institutions to Water Management in Arizona, California, and Colorado*. Washington, DC: Resources for the Future.

BROMLEY, D.W. 1982. Land and water problems: an institutional perspective. *American Journal of Agricultural Economics*, 64, 834–44.

BROMLEY, D.W. 1989. *Economic Interests and Institutions: The Conceptual Foundations of Public Policy*. Oxford: Basil Blackwell.

BROWN, F.L., MARS, C., MINNIS, M., SMASAL, S.A., KENNEDY, D. and URBAN, J.A. 1992. *Transfers of Water Use in New Mexico*. Albuquerque: New Mexico Water Resources Research Institute.

CAREY, J.M. and SUNDING, D.L. 2001. Emerging markets in water: a comparative institutional analysis of the Central Valley and Colorado–Big Thompson Projects. *Natural Resources Journal*, 41, 283–328.

CAREY, J., SUNDING, D.L. and ZILBERMAN, D. 2002. Transaction costs and trading behavior in an immature water market. *Environment and Development Economics*, 7, 733–50.

CHALLEN, R. 2000. *Institutions, Transaction Costs and Environmental*

Policy. Cheltenham, UK, and Northampton, MA, USA: Edward Elgar Publishing.
CHRISTIAN-SMITH, J., LEVY, M. and GLEICK, P. 2014. Maladaptation to drought: a case report from California, USA. *Sustainability Science*, 10, 1–11.
COASE, R.H. 1960. The problem of social cost. *Journal of Law and Economics*, 3, 1–44.
COASE, R.H. 1988. *The Firm, the Market, and the Law*. Chicago, IL: University of Chicago Press.
COLBY, B.G. 1988. Economic impacts of water law–state law and water market development in the Southwest. *Natural Resources Journal*, 28, 721.
COLBY, B.G. 1990a. Enhancing instream flow benefits in an era of water marketing. *Water Resources Research*, 26, 1113–20.
COLBY, B.G. 1990b. Transactions costs and efficiency in Western water allocation. *American Journal of Agricultural Economics*, 72, 1184–92.
COLE, D.H. 2002. *Pollution and Property: Comparing Ownership Institutions for Environmental Protection*. Cambridge: Cambridge University Press.
COMMONS, J.R. 1931. Institutional economics. *American Economic Review*, 21, 648–57.
CONNELL, D. 2007. *Water Politics in the Murray–Darling Basin*. Annandale, NSW: Federation Press.
CRASE, L., O'KEEFE, S. and DOLLERY, B. 2013. Talk is cheap, or is it? The cost of consulting about uncertain reallocation of water in the Murray–Darling Basin, Australia. *Ecological Economics*, 88, 206–13.
CRASE, L., O'REILLY, L. and DOLLERY, B. 2000. Water markets as a vehicle for water reform: the case of New South Wales. *Australian Journal of Agricultural and Resource Economics*, 44, 299–321.
EASTER, K.W. and MCCANN, L.M. 2010. Nested institutions and the need to improve international water institutions. *Water Policy*, 12, 500–516.
ERFANI, T., BINIONS, O. and HAROU, J.J. 2014. Simulating water markets with transaction costs. *Water Resources Research*, 50, 4726–45.
GARRICK, D. and AYLWARD, B. 2012. Transaction costs and institutional performance in market-based environmental water allocation. *Land Economics*, 88, 536–60.
GARRICK, D., MCCANN, L. and PANNELL, D.J. 2013a. Transaction costs and environmental policy: taking stock, looking forward. *Ecological Economics*, 88, 182–4.
GARRICK, D., SIEBENTRITT, M., AYLWARD, B., BAUER, C. and PURKEY, A. 2009. Water markets and freshwater ecosystem services: policy reform and implementation in the Columbia and Murray–Darling Basins. *Ecological Economics*, 69, 366–79.
GARRICK, D., WHITTEN, S. and COGGAN, A. 2013b. Understanding the evolution and performance of water markets and allocation policy: a transaction costs analysis framework. *Ecological Economics*, 88, 195–205.
GARRIDO, A. 1998. An economic analysis of water markets within the Spanish agricultural sector: can they provide substantial benefits? In: EASTER, K.W., ROSEGRANT, M. and DINAR, A. (eds), *Markets for Water: Potential and Performance*. New York: Kluwer.
GRAFTON, R.Q., LIBECAP, G.D., EDWARDS, E.C. and LANDRY, C. 2012.

Comparative assessment of water markets: insights from the Murray–Darling Basin of Australia and the Western USA. *Water Policy*, 14, 175–93.
HAGEDORN, K. 2008. Particular requirements for institutional analysis in nature-related sectors. *European Review of Agricultural Economics*, 35, 357–84.
HANEMANN, W.M. 2006. The economic conception of water. In: ROGERS, P., LLAMAS, M.R. and CORTINA, L.M. (eds), *Water Crisis: Myth or Reality*. CRC Press.
HARDIN, G. 1968. The tragedy of the commons. *Science*, 162, 1243–8.
HARRIS, E. 2011. The impact of institutional path dependence on water market efficiency in Victoria, Australia. *Water Resources Management*, 25, 4069–80.
HARRISON, K. 2013. Federalism and climate policy innovation: a critical reassessment. *Canadian Public Policy*, 39, S95–S108.
HEARNE, R.R. and EASTER, K.W. 1995. *Water Allocation and Water Markets: An Analysis of Gains-from-Trade in Chile*. Washington, DC: World Bank Publications.
HEARNE, R.R. and EASTER, K.W. 1997. The economic and financial gains from water markets in Chile. *Agricultural Economics*, 15, 187–99.
HEINMILLER, B.T. 2009. Path dependency and collective action in common pool governance. *International Journal of the Commons*, 3. Available at: http://www.thecommonsjournal.org/index.php/ijc/article/viewArticle/URN%3ANBN%3ANL%3AUI%3A10-1-100054/49.
HOWE, C.W. 2000. Protecting public values in a water market setting: improving water markets to increase economic efficiency and equity. *University of Denver Water Law Review*, 3(2), 357.
HOWE, C.W. and GOEMANS, C. 2003. Water transfers and their impacts: lessons from three Colorado water markets. *Journal of the American Water Resources Association*, 39(5), 1055–65.
HOWE, C.W., BOGGS, C. and BUTLER, P. 1990. Transactions costs as determinants of water transfers. *University of Colorado Law Review*, 61(2), 393–405.
HOWITT, R.E. 1994. Empirical analysis of water market institutions: the 1991 California water market. *Resource and Energy Economics*, 16, 357−71.
KIMMICH, C. 2013. Linking action situations: coordination, conflicts, and evolution in electricity provision for irrigation in Andhra Pradesh, India. *Ecological Economics*, 90, 150–58.
KISER, L.L. and OSTROM, E. 1982. The three worlds of action: a metatheoretical synthesis of institutional approaches. In: OSTROM, E. (ed.), *Strategies of Political Inquiry*. Beverly Hills, CA: Sage.
KRUTILLA, K. and ALEXEEV, A. 2014. The political transaction costs and uncertainties of establishing environmental rights. *Ecological Economics*, 107, 299–309.
KRUTILLA, K. and KRAUSE, R. 2010. Transaction costs and environmental policy: an assessment framework and literature review. *International Review of Environmental and Resource Economics*, 4, 261–354.
LIBECAP, G.D. 2005. The problem of water. Workshop on New Institutional Economics and Environmental Issues, INRA-ENESAD CESAER.
LIBECAP, G.D. 2009. Chinatown revisited: Owens Valley and Los Angeles – bargaining costs and fairness perceptions of the first major water rights exchange. *Journal of Law, Economics, and Organization*, 25, 311–38.
LIBECAP, G.D. 2011a. Institutional path dependence in climate adaptation:

Coman's 'Some unsettled problems of irrigation'. *American Economic Review*, 101, 64–80.
LIBECAP, G.D. 2011b. Water rights and markets in the US semiarid West: efficiency and equity issues. In: COLE, D.H. and OSTROM, E. (eds), *Property in Land and Other Resources*. Cambridge, MA: Lincoln Institute of Land Policy.
LIVINGSTON, M.L. 2005. Evaluating changes in water institutions: methodological issues at the micro and meso levels. *Water Policy*, 7, 21–34.
LUND, J.R. 1993. Transaction risk versus transaction costs in water transfers. *Water Resources Research*, 29, 3103–7.
MACDONNELL, L.J. 1990. *The Water Transfer Process as a Management Option for Meeting Changing Water Demands*. Boulder, CO: Natural Resources Law Center, University of Colorado.
MARSHALL, G. 2005. *Economics for Collaborative Environmental Management: Regenerating the Commons*. London: Earthscan.
MARSHALL, G.R. 2013. Transaction costs, collective action and adaptation in managing complex social–ecological systems. *Ecological Economics*, 88, 185–94.
MARSHALL, G., CONNELL, D. and TAYLOR, B. 2013. Australia's Murray–Darling Basin: a century of polycentric experiments in cross-border integration of water resources management. *International Journal of Water Governance*, 1, 197–218.
MCCANN, L. 2009. Transaction costs of environmental policies and returns to scale: the case of comprehensive nutrient management plans. *Review of Agricultural Economics*, 31, 561–73.
MCCANN, L. 2013. Transaction costs and environmental policy design. *Ecological Economics*, 88, 253–62.
MCCANN, L. and EASTER, K.W. 1999. Evaluating transaction costs of nonpoint source pollution policies. *Land Economics*, 75, 402–14.
MCCANN, L. and EASTER, K.W. 2004. A framework for estimating the transaction costs of alternative mechanisms for water exchange and allocation. *Water Resources Research*, 40(9), 1–6.
MCCANN, L., COLBY, B.G., EASTER, W.K., KASTERINE, A. and KUPERAN, K.V. 2005. Transaction cost measurement for evaluating environmental policies. *Ecological Economics*, 52, 527–42.
MEINZEN-DICK, R. 2007. Beyond panaceas in water institutions. *Proceedings of the National Academy of Sciences USA*, 104, 15200–15205.
MEINZEN-DICK, R. 2014. Property rights and sustainable irrigation: a developing country perspective. *Agricultural Water Management*, 145, 23–41.
MOLLE, F., WESTER, P. and HIRSCH, P. 2010. River basin closure: processes, implications and responses. *Agricultural Water Management*, 97, 569–77.
NORTH, D.C. 1990. *Institutions, Institutional Change and Economic Performance*. Cambridge: Cambridge University Press.
NORTH, D.C. 1994. Economic performance through time. *American Economic Review*, 84, 359–68.
NORTH, D.C. 2006. *Understanding the Process of Economic Change*. Princeton, NJ: Academic Foundation.
NUNN, S.C. 1990. Transfers of New Mexico water: a survey of changes in place and/or purpose of use, 1975–87. *Proceedings: 34th Annual New Mexico Water Conference, Water Resources Research Institute Report*, No. 248.

OSTROM, E. 1965. Public entrepreneurship: a case study in ground water basin management. Doctoral dissertation, University of California – Los Angeles.
OSTROM, E. 1990. *Governing the Commons: The Evolution of Institutions for Collective Action*. Cambridge: Cambridge University Press.
OSTROM, E. 2009a. A general framework for analyzing sustainability of social-ecological systems. *Science*, 325, 419–22.
OSTROM, E. 2009b. *Understanding Institutional Diversity*. Washington, DC: Princeton University Press.
OSTROM, E., BURGER, J., FIELD, C.B., NORGAARD, R.B. and POLICANSKY, D. 1999. Revisiting the commons: local lessons, global challenges. *Science*, 284, 278–82.
PETERSON, J.M., SMITH, C.M., LEATHERMAN, J.C., HENDRICKS, N.P. and FOX, J.A. 2011. The role of contract attributes in purchasing environmental services from landowners. Manhattan, KS: Kansas State University.
PUJOL, J., RAGGI, M. and VIAGGI, D. 2006. The potential impact of markets for irrigation water in Italy and Spain: a comparison of two study areas. *Australian Journal of Agricultural and Resource Economics*, 50, 361–80.
RAYMOND, L.S. 2003. *Private Rights in Public Resources: Equity and Property Allocation in Market-based Environmental Policy*. Washington, DC: Resources for the Future.
RITTEL, H.W. and WEBBER, M.M. 1973. 2.3 Planning problems are wicked. *Polity*, 4, 155–69.
ROSE, C.M. 1990. Energy and efficiency in the realignment of common-law water rights. *Journal of Legal Studies*, 19, 261–96.
ROSS, A. 2012. Easy to say, hard to do: integrated surface water and groundwater management in the Murray–Darling Basin. *Water Policy*, 14, 709–24.
RUML, C.C. 2005. The Coase theorem and Western US appropriative water rights. *Natural Resources Journal*, 45, 169.
SALETH, R.M. and DINAR, A. 2004. *The Institutional Economics of Water: A Cross-Country Analysis of Institutions and Performance*. Cheltenham, UK and Northampton, MA, USA: Edward Elgar Publishing.
SCHLAGER, E. 2005. Getting the relationships right in water property right. In: BRUNS, B.R., RINGLER, C. and MEINZEN-DICK, R.S. (eds), *Water Rights Reform: Lessons for Institutional Design*. Washington, DC: International Food Policy Research Institute.
SCHLAGER, E. and BLOMQUIST, W.A. 2008. *Embracing Watershed Politics*. Boulder, CO: University Press of Colorado.
SCHLAGER, E. and OSTROM, E. 1992. Property-rights regimes and natural resources: A conceptual analysis. *Land Economics*, 68.
SCHORR, D. 2012. *The Colorado Doctrine*. New Haven, CT: Yale University Press.
SKURRAY, J.H., PANDIT, R. and PANNELL, D.J. 2013. Institutional impediments to groundwater trading: the case of the Gnangara groundwater system of Western Australia. *Journal of Environmental Planning and Management*, 56, 1046–72.
SPEELMAN, S., FAROLFI, S., FRIJA, A. and VAN HUYLENBROECK, G. 2010. Valuing improvements in the water rights system in South Africa:

a contingent ranking approach. *JAWRA Journal of the American Water Resources Association*, 46, 1133–44.
SPROULE-JONES, M. 2002. *Restoration of the Great Lakes: Promises, Practices, and Performances*. Vancouver: UBC Press.
THOMPSON JR, B.H. 1993. Institutional perspectives on water policy and markets. *California Law Review*, 671–764.
WALLIS, J.J. and NORTH, D. 1986. Measuring the transaction sector in the American economy, 1870–1970. In: ENGERMAN, S. and GALLMAN, R. (eds), *Long-Term Factors in American Economic Growth*. Chicago, IL: University of Chicago Press.
WESTERN GOVERNORS ASSOCIATION (WGA). 2012. *Water Transfers in the West: Projects, Trends, and Leading Practices in Voluntary Water Trading*. Denver, CO: WGA.
WILLIAMSON, O.E. 1985. *The Economic Institutions of Capitalism*. New York: Simon & Schuster.
WILLIAMSON, O.E. 1998. Transaction cost economics: how it works; where it is headed. *De economist*, 146, 23–58.
YOUNG, R.A. 1986. Why are there so few transactions among water users? *American Journal of Agricultural Economics*, 68, 1143–51.
ZETLAND, D. 2011. *The End of Abundance: Economic Solutions to Water Scarcity*. Amsterdam: Aguanomics Press.
ZHANG, J., ZHANG, F., ZHANG, L. and WANG, W. 2009. Transaction costs in water markets in the Heihe River Basin in Northwest China. *International Journal of Water Resources Development*, 25(1), 95–105.

3. Unlocking the past: path dependency and intertemporal costs

> [H]ere is the mighty river and its tributaries, as yet largely undeveloped, affording possibilities of extensive use for water power in its many canyons and for irrigation in its desert valleys, which need only the life-giving water to make them productive and valuable . . . These rivers make possible not only the construction of large irrigation systems and the growth of towns, cities, and prosperous agricultural communities but also the generation of hydroelectric power for lighting, heating, industrial uses, and the transportation of freight and passengers. (Grover, in La Rue, 1916: 9–10)

> For nearly a century, the water of the Colorado has been at the center of an intense political, legal, and economic tug-of-war between agricultural, municipal, industrial, tribal, environmental, state, and federal interests . . . every drop of the Colorado River is carefully planned and controlled, taking on new obligations and owners at each of its thousands of headgates, dams, and diversions. By the time it crosses the border, the mighty Colorado is exhausted to little more than a large stream – a stream that in most years disappears long before it reaches its terminus, diverted for irrigation in Mexico. [Shortly after long-range supply and demand intersected for the first time in the Colorado River]. (Culp, 2001: 1)

INTRODUCTION

Rivers do not run dry overnight. The closure of the iconic rivers of the Western US and Southeast Australia has its roots in the ideas and incentives that shaped rural development in the late nineteenth and early twentieth centuries. The unintended consequences of these past decisions and technologies are hard to unravel because of path dependency: the simple notion that history matters. Past institutional choices and technologies opened some options for water resource development and allocation, and foreclosed others. Path dependency provides a lens for understanding the direction and pace of institutional change in water allocation reforms. On the one hand, path dependency reproduces the status quo due to resistance to change from vested interests, pervasive externalities stemming from the connectivity of hydrological and institutional systems, sunk costs of capital-intensive water infrastructure and the associated problem of stranded assets and, not least, distributional conflicts over property rights which create contractual obligations nested across multiple levels of collective action. At the same time, inertia is not inevitable. The simple message of this chapter is that water allocation reform does not occur in an institutional vacuum, but is influenced by historic trade-offs that bind the past with the future through choices of institutional design, processes of technological innovation, and the sequencing and interplay of the two. On its own, path dependency has been critiqued as descriptive and deterministic, producing 'Just So' stories that smack of inevitability with limited theoretical significance or policy relevance. Linking path dependency with transaction costs economics and theories of institutional change, however, enriches our understanding of the historical circumstances shaping contemporary reforms, as well as the implications of today's choices for future adaptation and adaptive efficiency.

Water allocation institutions in the three basins (the Murray–Darling, the Columbia and the Colorado) share a common law heritage. Despite similar starting points, water allocation institutions have followed divergent paths in the development of water rights systems, irrigation supply organizations, and the constitutional-level rules used to establish interstate apportionment agreements and river basin governance arrangements. In this chapter, I examine the sources and impacts of path dependency during the river basin development era, and the ongoing period of cap-and-trade water allocation reform. The mounting pressures of drought and development have opened policy windows for water allocation reform that can either unlock path dependency or be captured by vested interests. These river basin trajectories illuminate the role of 'intertemporal' costs: the costs associated with a loss of institutional flexibility due to historic choices. Unlocking the past therefore requires water rights reforms and river basin governance arrangements that preserve flexibility rather than replace today's constraints with tomorrow's albatross.

WATER ALLOCATION REFORM AT A CROSSROADS: THREE RIVERS SHIFTING COURSE

How has the past shaped prospects for water allocation reform, and what are the implications for institutional design, technological innovation and adaptive efficiency in cap-and-trade reforms? Two recent milestones in the multi-decadal water allocation reforms in the Colorado and Murray–Darling offer a useful entry point. In the space of two days in late November 2012, an international agreement was finalized between the US and Mexico on the Colorado (Minute 319), and the Australian Parliament adopted a statutory Murray–Darling Basin Plan.

These milestones illustrate two rivers shifting course. In the Colorado River, negotiators from the US and Mexico agreed to Minute 319[1] of the International Treaty on the Utilization of Waters of the Colorado and Tijuana Rivers and of the Rio Grande. Minute 319 created a basket of benefits attractive for both countries and the diverse interests within them. In the US, a primary purpose of Minute 319 was to include Mexico in shortages triggered under the interim guidelines governing the lower basin within the US since 2007. In exchange, Mexico became eligible for new water banking mechanisms to store unused annual entitlements in Lake Mead – the lower basin's main storage reservoir – to buffer impacts of prolonged droughts. Mexico also received critical capital infusions to repair irrigation and water distribution infrastructure damaged by an earthquake in 2010.

The Minute also enabled the reallocation of water entitlements to restore the Colorado Delta. A regional coalition of non-governmental organizations (NGOs) collaborated with Mexican irrigators to reallocate water for minimum flows (known as 'baseflows') via market-based water rights acquisitions. However, deltas also require periodic pulses of water to mimic the peak flows of the hydrograph and restore floodplain wetlands that harbour migratory birds. Pulse flows require even more coordination than baseflows because they depend on operational changes in upstream storages. The periodic pulse flows are being coordinated by non-profit environmental organizations, and state, international and federal agencies. The first pulse flows were delivered in late March 2014 amidst sustained drought in the Colorado Basin and the state of California (Flessa et al., 2013).

Although only a five-year trial, the agreement signalled the reopening of a closed river over 40 years after Lake Powell – the basin's upstream reservoir – began filling in the late 1960s. It becomes even more remarkable when considering this milestone from the vantage point of 2004, when shortage sharing discussions commenced in the Lower Colorado, as state governments sought alternatives to the acrimonious legal battles that dominated the interstate apportionment of water in the first half of the twentieth century. These milestones demonstrate how climate change and prolonged drought have the potential to disrupt prevailing water allocation institutions and to open opportunities for water allocation reform by overcoming the paralysis of vested interests and inflexible water rights systems (Benson, 2012). Minute 319 culminated a steady shift in the basin's trajectory over the preceding eight years, provoked by unprecedented drought conditions, which produced a period of unexpected cooperation. The Record of Decision announcing the interim shortage guidelines adopted in 2007 illustrated the promising directions in the Colorado's water allocation reforms (US Bureau of Reclamation, 2007):

> a unique and remarkable consensus emerged in the basin among stakeholders . . . to preserve flexibility to deal with further challenges such as climate change and deepening drought, implement operational rules for a long – but not permanent – period in order to gain valuable operating experience, and continue to have the federal government facilitate – but not dictate – informed decision-making in the Basin.

In other words, the Record of Decision embodied a set of principles for adopting institutional reforms to achieve adaptive efficiency in a context of uncertainty and path dependency. Although the challenges facing the Colorado River include a projected 4 billion m^3 average annual shortfall by 2060 under baseline conditions, the principles guiding these recent reforms show unexpected promise.

In the Murray–Darling Basin (MDB), the Australian Parliament adopted the Murray–Darling Basin Plan a mere two days later. The Plan defined sustainable diversion limits and basin-wide environmental water needs pursuant to the Commonwealth Water Act (2007). Contrary to the celebratory tone of Minute 319 and the broad-based support it engendered, the 2012 Basin Plan marked a fragile compromise amidst threats of High Court challenges by downstream South Australia (Webster and Williams, 2012) and deep-seated mistrust among conservation groups and irrigation communities in regional Australia. The Plan sought to fulfil the 2007 Act's objective to optimize ecological, social and economic outcomes of water use in the Basin, a comprehensive goal established during the throes of the Millennium Drought from 1997 to 2009. Like the Colorado agreement, the Basin Plan reopens a closed river by increasing the reliability of water reaching the Murray Mouth.

Negotiations over the Murray–Darling Basin Plan became acrimonious and triggered a backlash: first in regional agricultural communities, followed by interstate tensions, particularly in the state of South Australia as a vulnerable downstream state, and in New South Wales as an upstream state targeted for water buybacks for the environment. More than 12000 comments were submitted during the consultation process in 2011–12, including almost 5000 associated with organized campaigns. In one sense this public participation is a sign of healthy political contestation, but it needs to be viewed against the counterfactual. In contrast with the Colorado, this contentious turn was unexpected from the vantage of 2004 when the outlook for the MDB was considerably more optimistic. Contentious politics have marked even the most promising epochs of water allocation reform since the Murray–Darling Basin Initiative started in the mid-1980s, but 2004 was arguably a high-water mark in the multi-layered water allocation reforms to establish water markets and sustain the diverse public goods generated by the basin.

The National Water Initiative of 2004 was a systematic effort to harmonize cap-and-trade reforms across states and recover water for the environment. State governments in Southeast Australia and the Commonwealth government also adopted the 'Living Murray First Step',[2] which led to a 500 GL commitment to environmental water recovery and a return to sustainable levels of extraction coordinated at the river level for the Southern Murray. The Initiative harmonized water rights reforms adopted the previous decade and built on intergovernmental cooperation since the Murray–Darling Basin Agreement of the late 1980s. Drought deepened in 2006 and 2007 and created a window for reform that aligned with wider political agendas as the Howard government came to a close (Horne, 2013). The 2007 Water Act aimed to accelerate the reform process, and amendments to the Act in 2008 injected almost AU$13 billion to, *inter alia*, purchase water from irrigators for the environment and invest in irrigation infrastructure to conserve water as part of the Water For the Future program and its Restoring the Balance and Sustainable Rural Water Use and Infrastructure programs. In the process, the Act represented a constitutional-level rule change from a consensus to unilateral decision-rule, compromising the hard-fought framework for interstate cooperation that had coalesced over two decades (Connell, 2007).

In 2004, the outlook was for (continued) lock-in in the Colorado, hostage to its history of litigation, and for cautious optimism to scale up reforms in the Murray–Darling. The outlook for the Columbia Basin lies between the two poles, muddling through with pockets of reform taking root in sub-basins across the four US states. In 2003, there were nascent efforts to coordinate local implementation at the basin scale through the Columbia Basin Water Transactions Program and sub-basin planning efforts facilitated by the Northwest Power and Conservation Council. Flash forward ten years and the Colorado and Murray–Darling Basins shifted course. The Colorado is now poised to consolidate its gains, albeit slowly and steadily, and the Murray–Darling is hampered by polarized politics. Meanwhile, the Columbia has experienced isolated successes that have proven difficult to scale up. Why?

The central thesis of this chapter is that water allocation reforms do not occur in an institutional vacuum due to path dependency: the notion that historical decisions constrain subsequent choices. Positive feedbacks increase returns to existing trajectories and make reversals or switching courses costly (North, 2006). Path dependency shapes the possible trajectories of institutional change because interests and organizations created under prevailing institutional arrangements will resist changes that threaten their viability. Path dependency is therefore a key explanatory factor, frequently acknowledged but generally weakly integrated into policy analysis (Heinmiller, 2009; Pierson, 2000; Kay, 2005). However, path dependency is not synonymous with inertia. Instead, a focus on path dependency underscores the role of time and the dynamic interplay of stability and change.

Nobel laureate Douglass North (2006: 51) notes that path dependency has been 'used, misused and abused'. Kay (2005) critiques path dependency in policy analysis by identifying three deficiencies, namely that path dependency has often: (1) failed to explain the evolution of decision-making; (2) privileged stability over change; and (3) ignored normative implications of a focus on efficiency. Nevertheless, I argue that the benefits of a historical perspective outweigh the analytical weaknesses if the deficiencies are acknowledged and addressed. First, path dependency is not a theoretical variable; hence its lack of explanatory power can be remedied by linking path dependency with relevant theoretical constructs, in this case those examining the relationship between transaction costs and institutional change. Transaction costs concepts of lock-in, intertemporal trade-offs and switching costs can be used to analyze a sequence of interrelated decisions. Second, a historical perspective has intrinsic value because 'deprived of their roots, stories are sickly things' (Powell, 1989: 37). Moreover, the patterns of institutional change in the three basins can be understood by pinpointing critical junctures and pivotal events, which expose potential for periodic transformations that break through incremental changes to produce 'punctuated equilibrium' – a term borrowed from biological theory to capture long periods of political and institutional stability punctuated by sudden shifts (Baumgartner et al., 2009). Third, path dependency can be used to analyze the impact of past decisions on adaptive efficiency instead of an exclusive focus on stability. Finally, the normative focus on efficiency is questioned throughout this book by emphasizing adaptive efficiency in relation to other evaluative criteria shaping long-term institutional performance, such as equity, sustainability and accountability, as outlined in the previous chapter. The evolution of institutions governing water allocation through water trading and river basin governance cannot be understood without tracing the equity norms, issues of legitimacy and accountability considered here as part of a broadened conception of adaptive efficiency.

A focus on path dependency in water allocation reform is nothing new, however (for example, Ingram and Fraser, 2006; Libecap, 2011): past institutional and technological changes are perpetuated by those with an economic stake in maintaining the status quo, for example irrigation organizations, farmers and bureaucratic institutions. Path dependency is therefore interwoven with the distributional concerns of political economy. The costs of institutional change are concentrated, local and immediate, while the benefits are often diffuse, distributed and delayed (cf. Harrison, 2013). These past changes hold a 'lasting grip' until shocks enable periodic institutional transformations (Williamson, 1998). In the context of market-based water reforms, the politics and design of the cap, initial allocation and tradable water rights merit careful analysis of path dependency and intertemporal trade-offs in order to understand how past and current choices impact upon future flexibility and long-run adaptive efficiency.

In the rest of this chapter, I use path dependency and transaction costs economics to introduce the three river basins and their trajectories of water resource development. The chapter defines path dependency in the context of contemporary cap-and-trade water allocation reforms: establishing caps, the initial assignment of rights and adjusting trading rules. The chapter then retraces the roots of contemporary constraints by examining the relationship between path dependency and institutional change to uncover the lock-in effects and intertemporal trade-offs of past decisions. The following section provides a short history of the three river basins, to identify the intertemporal trade-offs of historic institutional choices and consider the implications for contemporary and future reforms. In so doing, I aim to 'restore a sense of the options that confronted people at the time, to show the grit and friction that was evident' in the contested political economy of river basin development and allocation institutions (Blackbourne, 2006: 12). This chapter foregrounds 'pivotal decisions and critical periods' shaping the trajectories of institutional change in water allocation (Powell, 1989: xiii). The concluding section illustrates the importance of flexibility and sequencing as an institutional design challenge and sets the context for the remaining chapters. It also highlights the interplay of institutional choices and infrastructure technologies.

PATH DEPENDENCY IN WATER ALLOCATION REFORM: SOURCES OF LOCK-IN

Institutions and technologies are self-perpetuating because constituencies who benefit from their persistence lobby to maintain them even in the face of a broad coalition for change. In transaction costs terms, the costs of reversing or changing past commitments are prohibitively high and can lead to lock-in. This is not inherently bad, as stability and security of property rights were used to encourage investment and initial economic development of the regions. Path dependency therefore captures the tension between opposing forces in water allocation reform: continuity and change, vested interests and emerging values, and stability and flexibility in property rights (Ingram and Fraser, 2006; Ostrom and Cole, 2010).

In closed river basins, path dependency means that water use patterns may become fixed in historic locations, principally for irrigation and often for crops and communities chosen for political and regional development purposes, not strictly for economic efficiency in welfare-maximizing terms. Institutions and infrastructure are designed to reduce uncertainty, and to cope with complex interactions of hydrology and society in regions prone to climate variability. These infrastructure systems have become intricate and complex. Small changes in water use patterns may have unpredictable social and political economic impacts.

In sum, path dependency 'mold[s] transaction costs, the extent of markets, and the nature of resource use, management, and investment' (Libecap, 2011: 76). Unintended consequences of past decisions can transform certainty into rigidity that impedes adaptation. Transaction costs are one symptom of that rigidity. Libecap (2011) notes that these choices were once considered institutional innovations to address seemingly intractable challenges of irrigation development. Yesterday's innovations have made it very costly to adapt to the pressures of today and tomorrow (Libecap, 2011).

Path dependency influences the evolution of water allocation institutions in the Western US and Southeast Australia in terms of at least three major choices: water rights systems, irrigation institutions and interstate apportionment agreements (and the associated river basin governance arrangements) (see Table 3.1). These choices include multiple levels of rules, including the operational, collective choice (decision-making) and constitutional-level actions and rules identified by Kiser and Ostrom (1982) and Ciriacy-Wantrup and Bishop (1975) (see Table 3.1).

Technological systems, formal institutions and norms interact in path dependent water resource development and allocation policy. Water is distinguished by its capital-intensive infrastructure for storage and distribution, which is lumpy and long-lived with a lifespan of three to five decades or longer (Hanemann, 2006). In this sense, irrigation infrastructure physically inscribes water access onto the landscape in ways that become difficult to change (Bruns and Meinzen-Dick, 2001). The Hoover (Colorado), Hume (Murray–Darling) and Grand Coulee (Columbia) are iconic dams positioning the regions as pioneers of government-led irrigation and hydropower systems with modern storage infrastructure

Table 3.1 Path dependency across institutional arrangements and levels of action

Institutional arrangement	***Level of action***	***Sources of path dependency***
Water rights:		
Rules governing access, use, and withdrawal	Operational	Requirement for a physical water diversion in the Western US
Rules governing management decisions	Collective choice	No-harm rule, preventing water rights transfers with unmitigated impacts
Irrigation supply organizations:		
Rules governing operations and maintenance of infrastructure	Operational	Exit fees as a surcharge on water sales leaving organization boundaries
Rules governing changes in ownership	Collective choice	Voting and membership rules restricting water sales
Interstate apportionment:		
Rules governing interstate water allocation	Operational	Proportional versus fixed allocation rule
Rules governing procedures to change allocation	Collective choice	Commission voting rules requiring consensus

(White, 1957; Worster, 1985; Reisner, 1986; Connell, 2007). The development paths in these basins have become international models, widely emulated and adapted for semi-arid regions with 'difficult hydrology' (Worster, 1985; Grey and Sadoff, 2007). The hydraulic era had environmental consequences, cost recovery challenges and financing issues that converged in the late 1970s and early 1980s to mark the end of the first wave of infrastructure development; recent reforms have included a resurgence of investment in infrastructure to modernize storage and distribution systems and build new ones, including aquifer storage and recovery systems.

Ingram and Fraser (2006) elaborate the policy and behavioral attributes of path dependency. Policy discourses create policy images or framings that become self-reinforcing through professional networks and vested interests. The Colorado, Columbia and Murray–Darling shared a common policy framing until the early 1980s premised on a goal of 'routinizing the irregular' patterns of streamflow variability – both seasonal and inter-annual – to support a rural development model for irrigation and hydropower production (Lach et al., 2005). The policy choices and technologies adopted during this period created communities which benefit from the maintenance of the status quo, namely irrigators, irrigation supply organizations, some cities and associated state and federal regulatory agencies. These interests coalesce into an 'iron triangle' of water users, engineers and bureaucrats (Molle and Wester, 2009; McCool, 1987). Professional bureaucracies form expert networks that administer the prevailing regulatory framework and infrastructure and become linchpins in the iron triangle. In the Colorado River and Western US, for example, water users, federal and state agency staffs have been described as 'water buffalo' due to their propensity to move slowly (Pittock, 2013; Waterman, 2010; McDermott, 1998). The contested political economy of water allocation stacks the deck against reallocation and institutional change: costs of change are concentrated on irrigation communities, and borne up-front, while the benefits are diffused and often deferred. Recalling the calculus of change posited by Basurto and Ostrom (2009) and described in Chapter 2, this distribution of benefits makes the coalition for change more difficult – and costly – to form. Politics, technology and the weight of history are therefore a fundamental constraint on market-enabling water allocation reforms.

In a comparison of the Colorado, Murray–Darling and Saskatchewan rivers, Heinmiller identified four sources of technological and institutional path dependency in cap-and-trade water reforms: vested interests, network effects, sunk costs and property rights governed by both formal and informal contracts (Heinmiller, 2009). Together these factors raise 'the difficulty and cost of reform . . . reinforcing the institutional status quo' (Heinmiller, 2009: 135). Below, I examine these four interacting sources of path dependency across multiple stages of reform in the Colorado, Columbia and Murray–Darling. Accounting for the combined impact of network effects (that is, pervasive externalities due to hydrological and institutional connectivity) and property rights (formal and informal contracts), Challen (2000: 150) notes:

> Transfers of property rights down an institutional hierarchy [that is, from states to individuals] can be difficult to reverse in so far as the transition costs of 'clawing back' the property rights at a later date are likely to be high. Consequently, devolution of property rights down the hierarchy may reduce the flexibility of the institutional structure with respect to future reforms.

Therein lies the conundrum for institutional change and adaptive efficiency in the water allocation reforms underway in the Colorado, Columbia and Murray–Darling, three closed or closing river basins. In these settings, water markets are being developed in a context of overallocation with intensified competition and potential for conflict between private consumptive uses (irrigation) and public non-consumptive uses (for example instream flows).

THE COSTS AND CONSEQUENCES OF LOCK-IN: IMPLICATIONS FOR INSTITUTIONAL DESIGN AND SEQUENCING

The foregoing demonstrates that historic decisions about water rights, irrigation organizations and interstate apportionment, as well as the associated technologies, affect the costs of future institutional transitions and switching costs. The transaction costs concepts of intertemporal trade-offs and lock-in have been developed to analyze the path dependent evolution of cap-and-trade water allocation reforms. Challen used the term 'intertemporal transaction costs' to describe the 'reduction in future flexibility of an institutional structure [that] may constitute a cost to society' (Challen, 2000: 150). Decreasing flexibility implies higher switching costs. Intertemporal transaction costs expose the trade-offs associated with a sequence of historic and current policy choices by clarifying the consequences of these choices for future institutional reform and adaptive efficiency (Challen, 2000; Marshall, 2013). Marshall (2005, 2013) described these intertemporal costs in terms of lock-in costs. The impacts of institutional and technological lock-in costs become increasingly important in a context of uncertainty, complexity and conflicts, which create the need for learning and flexibility to respond to shifting information and preferences. Lock-in costs represent a loss in 'quasi-option values', that is, the degrees of freedom available to society for future institutional changes (Marshall, 2005).

A transaction costs perspective on path dependency makes the costs of past decisions visible. Consider the effects of path dependency in terms of the calculus of transaction costs and institutional change described in the last chapter (Basurto and Ostrom, 2009). Embedded within the 'upfront costs' of institutional change are the costs associated with past decisions, which have created a stream of benefits for groups who will resist changes that undermine their vested interests, contributing to lock-in. These consequences of past decisions can be construed as a form of 'intertemporal costs' associated with a loss in flexibility (Challen, 2000; Marshall, 2005; Marshall et al., 2013). I propose a new category of costs, C1a, to capture the intertemporal costs imposed by past decisions which limit future flexibility.

$$D > (C1 + C1a + C2 + C3)$$

where:
D = benefits of rule change;
$C1$ = 'upfront costs' of developing and adopting new rules;
$C1a$ = 'intertemporal costs' imposed by past decisions which limit future flexibility;
$C2$ = 'short-term costs of implementing new rules';
$C3$ = 'long-term costs of monitoring and maintaining a self-governed system over time'. (Basurto and Ostrom, 2009)

The costs and benefits of institutional change must be considered at both the individual and collective levels. Path dependency affects the political coalition needed to achieve a change. Individuals who benefit from an institutional change (the winners) are forced to bargain with those bearing the concentrated costs of the change (losers).

The notion that reductions in future flexibility are a form of transaction cost begs fundamental questions about institutional design: flexibility compared to what? Assessing lock-in costs requires comparison of alternative institutional choices for: capping water diversions, the initial allocation and trading rules governing reallocation. Kay (2005) notes the difficulty of constructing meaningful counterfactuals under the political, historical and technological circumstances that prevailed at the time. However, it is feasible to compare institutional design options in terms of their effects on future flexibility.

Following Tietenberg (2002), the cap, or aggregate limit, for resource use and pollution is based on some conception of sustainable use. The key institutional design variables are: (1) whether in fact the cap is based on sustainable use versus historic use (which may already be unsustainable); and (2) whether the aggregate limit is fixed or can be reviewed periodically based on new information or sustainability values. These choices have important intertemporal implications (Table 3.2). The establishment of a

Table 3.2 Institutional design and intertemporal costs

	Intertemporal costs ($C1a$)	
Design elements	*Higher*	*Lower*
Cap	Historic use – 'grandfathering'	Adaptive – 'sustainable'
Allocation	Prior appropriation	Proportional shares
Trading	'No-harm' rule	Competition policy
Private property	Stronger claims	Weaker claims
Flexibility	Lower	Higher

cap often occurs after sustainable use levels have been exceeded. If a cap is based on historic uses, it converts tacit and informal claims into formal or more formal rights and expectations. Formal rights become costly to adjust if sustainability criteria later warrant a reduction in cumulative diversion levels (Heinmiller, 2006; Colby, 2000). The cap is therefore inextricably bound up with the initial formal allocation of rights.

The initial allocation is therefore a critical institutional choice (Krutilla and Krause, 2010). There are four broad options for the initial allocation: random access (lotteries); first come, first served; administrative rules based on eligibility criteria; or auctions (Tietenberg, 2002). Ostrom and Basurto (2010) demonstrate a number of additional permutations for irrigation water systems based on seven types of rules, including the boundary, position and scope rules governing access and decision-making about irrigation infrastructure and access, making it necessary to consider a much larger number of possible configurations. In terms of lock-in costs, the initial assignment of property rights has considerable impact on trajectories of institutional change (Heinmiller, 2006; Heinmiller, 2009; Krutilla and Krause, 2010; Coase, 2012). 'Grandfathering' is a common basis for defining property rights because it recognizes historic investments and is therefore the method most likely to gain political approval (Tietenberg, 2002). However, grandfathering formalizes these claims and establishes expectations of secure water rights that can become more costly to alter than other standards used for defining property rights (Raymond, 2003). An appreciation of path dependency and the associated intertemporal trade-offs led Challen (2000) to highlight the risks associated with allocation decisions that reassign water rights from state ownership to private control, as private property rights become more difficult and costly to adjust as preferences change.

Finally, trading rules, or reallocation options, can be designed along the spectrum from principles of free competition to stringent protection for third parties; water trading rules have erred toward the latter, particularly in the US, by preventing harm (see below) and reinforcing irrigation district control by restricting water trade out of districts. Such rules may combine with infrastructure improvement schemes that reinforce lock-in by exacerbating the problem of stranded assets. Together, these institutional choices about the cap, initial allocation and trading rules create a spectrum from flexible to more rigid, in the extreme case raising the prospect of an 'anti-commons' in which formal specification of rights contributes to paralysis and underuse, rather than overuse (Heller, 1998).

These institutional design choices are sequential. Cap-and-trade water allocation reforms involve multiple stages. These stages become linked by path dependency because decisions during initial stages impose deferred costs. Colby (2000) identifies five stages of cap-and-trade reforms:

(1) resource abundance with informal rights and rules; (2) perceived scarcity and conflict; (3) a period of political disputes regarding capping, allocation and trading regulation; (4) the establishment of the cap, allocation of rights and trading rules; and (5) expansion of trading activity. Considered in terms of transaction costs and water allocation reform, stage 3 can involve high institutional transition costs[3] due to the distributive politics and the race to legitimize informal claims during the grandfathering phase (Heinmiller, 2006). Stage 4 is marked by high static transaction costs[4] during early trading to interpret and implement cap, allocation and trading regulations. In some situations, the political and institutional hurdles of stage 4 are too high, particularly for complex economic goods with interdependent private and public values, such as water infrastructure and environmental flows. However, when initial impediments have been overcome, the expansion of trading activity is a sign that benefits of reallocation have increased, or static transaction costs have decreased, or, most likely, some combination, to realize the perceived gains from trade. Thus, stage 5 is typically marked by relatively lower transaction costs even though the need for periodic adaptation and institutional reform remains (see Figure 3.1), as exemplified

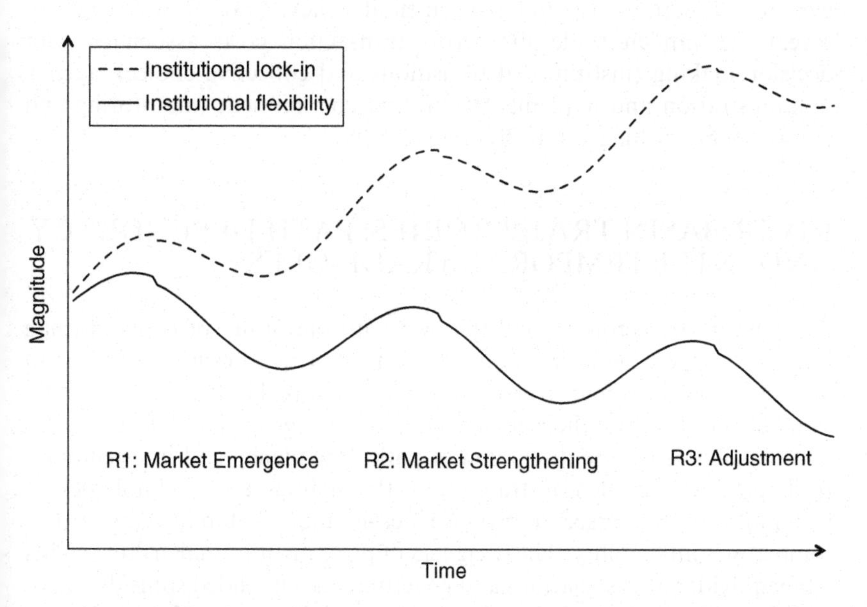

Note: The divergence of trajectories can be traced to historic institutional choices which altered the range of future options, for example grandfathering versus renewable permits.

Figure 3.1 Lock-in costs and intertemporal trade-offs in cap-and-trade water allocation reform

by the comprehensive basin planning effort used to revise the cap on water diversions in the Murray–Darling.

Trajectories of cap-and-trade reform processes are not linear. Efforts to decrease transaction costs in early phases of reform (for example, grandfathering) may prove short-sighted by shutting out key stakeholders from the process of developing the cap, allocation or trading processes. Such efforts can prove self-defeating over the long term and dampen political will for market-based reallocation because 'costless trading will produce inefficiencies when externalities are produced, or public goods are diminished' (Colby, 2000: 652). Cap-and-trade water allocation reform must be considered in this historical and political economic context to achieve adaptive efficiency over the long term. This is captured best by Colby (2000: 652): 'Transaction costs resulting from legitimate oversight are a necessary part of implementing an efficient market. However, public agency oversight . . . also gives affected stakeholders who are averse to cap-and-trade policies a means to express their objectives. Stakeholders . . . can impose costs and delays, a crucial form of bargaining power'.

McCann et al. (2005) highlight the need for typologies and chronologies of transaction costs in environmental policy. Trajectories of reform reveal the implicit trade-offs across transaction costs associated with decision-making (institutional transitions and collective choice), agency (administration and implementation) and commitment (monitoring and enforcement) (Schlager and Blomquist, 2008).

RIVER BASIN TRAJECTORIES: PATH DEPENDENCY AND INTERTEMPORAL TRADE-OFFS

River basin 'trajectories' bind the past with the future of cap-and-trade reforms in the Colorado, Columbia and Murray–Darling. Molle and Wester (2009: 1) define river basin trajectories as the 'long-term interactions of societies and their environments' in water resource development and management. The discussion of path dependency and intertemporal trade-offs (above) demonstrates how the politics and technologies of water allocation intersect to produce lock-in: today's stability can become tomorrow's inflexibility. The designers of prior institutional arrangements – the individuals, associations and governments – intended some degree of lock-in and path dependency, that is, they sought stability for investment and security. In other instances, lock-in causes unintended consequences and paralysis that benefits a narrow group of vested interests. The benefits of stability versus costs of inflexibility can become visible through comparative and longitudinal perspectives, tracing river basin development and institutional reform from the vantage of path dependency and intertemporal trade-offs tied to key decisions and technologies. This provides a foundation for the forthcoming chapters on market-based water allocation reform in a transaction costs world.

Here I trace three phases of the river basin trajectories during the water resource development era: irrigation settlement, interstate apportionment and the initial phase of cap-and-trade reforms. I consider key institutional design choices in these periods in light of the options debated at the time, and compare the three basins in terms of lock-in. The chapter finishes with a brief comparison of the 2004 and 2014 outlooks for the three basins. The Colorado River's experience during that ten-year period is used to illustrate the potential for transformation during crisis but also the risk of maladaptation, given the lessons from the past. This historical perspective sets the stage, or initial conditions, for the in-depth analysis of institutional design and performance at the intersection of water trading, environmental flow restoration and river basin governance arrangements in an emerging (Columbia) and maturing (Murray–Darling) market-based water allocation reform process in Chapters 4 and 5, respectively.

The physical and human geography of the three basins is a starting point to trace the interactions between people, institutions and environment in water allocation reform, following the introduction of basin closure processes in the three rivers presented in Chapter 1. All three rivers are large and complex systems, spanning multiple political jurisdictions and diverse interests and uses. The overarching similarities are semi-arid conditions, climate variability, overallocation and therefore some form of river basin closure, either at the tributary level as in the Columbia, or at basin-wide levels in the Colorado and Murray–Darling (Kenney et al., 2009; Turral et al., 2009; Aguilera-Klink et al., 2000). Despite these broad similarities, there are several important contrasts and critical junctures when decisions during crises and policy windows have led to divergence in the trajectories and property rights systems, particularly in the application of common law principles and different sets of constitutional rules to water development and allocation in multi-jurisdictional rivers.

An Era of Abundance? The 'Unsettled Problem' of Irrigation

Long before contemporary conflicts and competition for water, there was a perception of relative abundance; particularly if irrigation and hydropower development could unlock and leverage the vast resources of the regions (as noted by Grover in the opening quotation for the chapter). Today there are many ways to qualify and quantify water availability: seasonal and annual precipitation and runoff, per capita availability of renewable freshwater (Falkenmark, 1989), the ratio of water withdrawals to availability (Vorosmarty et al., 2005), and storage per capita. In physical geographic terms, the Colorado and Murray–Darling Basins, as well as the interior tributaries of the Columbia, are semi-arid to arid regions with a snowmelt-driven hydrograph; the Colorado and Murray–Darling fall within the lowest quartile for mean annual runoff among river basins globally, and also the highest quartile in terms of inter-annual variability associated with prolonged droughts (Garrick et al., 2013). Seasonal variability of supply and demand creates a mismatch between peak runoff of snowmelt in the spring and peak demand in late summer for irrigation during the natural low flow in the hydrograph. This may lead to dewatered rivers and tributaries, dependence on groundwater and/or unsustainable extraction levels that encroach on environmental flows.

Seasonal variability is compounded by inter-annual and decadal fluctuation in water availability. Large storage infrastructure has been constructed to cope with extremes and contribute to regional development through irrigation and hydropower production. Supply-side proposals to buffer variability and address growing demand remain a focus of large importation (interbasin transfer) schemes today. Inter-basin transfers are likely to remain a prominent option in the future due to their political attractiveness, that is, distributed costs, with often concentrated benefits for irrigators and cities (US Bureau of Reclamation, 2012; Gleick et al., 2014). Finally, spatial variability compounded the challenges of water resource development and allocation; there is a mismatch between the demand and supply. Mining and agriculture required diversions to deliver water to lands often at long distance from the stream channel. Moreover, the snowmelt-driven hydrograph means that the fertile river valleys of the lower basins are far removed from the headwaters regions where the vast majority of runoff is generated and stored. In the Colorado River, for example, approximately 85 per cent of runoff is generated upstream of Lees Ferry (Ficklin et al., 2013), while the arable land is concentrated in the lower portion of the basin and in valleys lying outside of the basin boundaries. In this context, the early twentieth century was associated with a lack of infrastructure and institutions to harness available water supplies at the times and places needed for productive uses such as irrigation, mining and hydropower. There was never an era of abundance per se for early European settlements of Western North America and South east Australia, but rather a period of development in a harsh and variable climate.

The classification of rivers in terms of scarcity or abundance is therefore in large part a social construction (Aguilera-Klink et al., 2000). Development of the three basins was driven by what Coman (2011 [1911]) described as the 'unsettled problem' of irrigation as settlers shifted from boom-and-bust mining to government-supported smallholder irrigation agriculture as a regional development and settlement policy in the second half of the nineteenth century (Libecap, 2011; Ostrom, 2011). Fulfilling the irrigation potential of the regions involved water resources development through property rights institutions, innovations in irrigation technology and organizational development and, not least, subsidies by national and colonial governments to finance capital-intensive infrastructure.

This depended in turn on the interjurisdictional agreements and river basin institutions to share costs and benefits. Settlers appropriated and diverted water from streams governed by a hybrid of common law principles and ad hoc rules and norms adapted to a territory with physical conditions far removed from the comparably tame hydrology of colonial centers in the Eastern US and coastal settlements of Australia. Private entrepreneurs and associations invested in capital infrastructure for storage and distribution, and larger national and subnational governments ultimately passed legislation and granted subsidies and loans, particularly when the investments by irrigation associations failed to achieve the economies of scale and sustainability required for irrigation societies in these semi-arid regions.

Colorado

The Colorado experienced a long and ornate history of exploration and exploitation marked by 'romantic incidents' replete with 'discoveries, starvations, battles, massacres, and lonely, dangerous journeys' (La Rue, 1916: 16). Kenney (2009) describes this era of (relative) abundance as one of exploration and settlement in the Colorado River, dominated by the 'private commodity paradigm' of water development[5] highlighted by the rush for gold and mineral deposits, first by Spanish *conquistadors* and later by the French and British.[6] When the mining boom went bust, the same profit motives were redirected to early irrigation schemes. The infamous California Development Company diversion developed Imperial Valley before heavy flooding overwhelmed the Alamo Canal, formed the Salton Sea, and bankrupted the company. The Imperial Irrigation District petitioned for federal investment to control floods and re-engineer the canal, setting off a pattern of federal response to powerful local irrigation organizations and their lobbying. It also illustrated the power of crises in unlocking path dependent reform to settle debates over competing institutional alternatives, which at the time centered on the public versus private provision of irrigation schemes.

Murray–Darling

Similar entrepreneurial forces drove initial settlers in Victorian gold mining expeditions in the Murray–Darling after British settlers navigated upstream from the Murray Mouth in search of riches. Like the Western US, the era of early water resource development did not begin with a blank slate, despite the observations of early European explorers. Aboriginal communities depended on water for 'incipient farming' and participated in 'quasi-engineering' (Powell, 1989). European settlement of the Murray followed exploration by Charles Sturt in the 1830s and a period of local and private irrigation organizations succeeded by government-led irrigation development after the 1886 Victoria Water Act. These similarities were not accidental, but rather the product of deliberate efforts by Australian architects to emulate the successes – and avoid the failures – of incipient irrigation economies in the Western US and India (Deakin, 1885).

Columbia

The particular economic drivers in the Columbia Basin differed, where fur trade and salmon canneries dominated initially (Lichatowich, 2001).[7] For this reason, Wandschneider (1986) notes that an era of abundance in the Columbia Basin prevailed until the 1970s, which did 'wonders for social harmony'. The Hudson Bay Trading company, originated by England in the seventeenth century, played an entrepreneurial role in the Columbia, also aided by generous concessions and infrastructure development by early territorial governments and their support from the federal government.

Resource extraction and large capital infrastructure investments as paths to economic development were the common thread across the basins in this era. The cyclical rise and subsequent decline in mining industries (and the fur trade of the Columbia), elevated the importance of irrigated agriculture as part of a conscious national development agenda of colonial settlement and expansion; the decisions had clear benefits, albeit often concentrated and overstated, but, equally, lock-in consequences that were unanticipated or underestimated at the time.

In short, with the comfortable vantage of hindsight, the benefits delivered by colonial era development can be weighed against upfront (*C1*) transaction costs and the deferred, or intertemporal (*C1a*), transaction costs imposed on future institutional transitions. The legacy of these intertemporal trade-offs (*C1a*) can be traced to three choices from this era that exert continued influence on contemporary trajectories: (1) water rights systems; (2) irrigation supply organizations; and (3) large storage infrastructure and the accompanying national, interstate and river basin governance arrangements devised to enable such large and long-lived investments.

Comparison: common law, diverging water allocation institutions

By the 1850s in the Colorado River, the limits of the riparian doctrine were being recognized. A landmark Colorado Supreme Court decision, *Coffin v. Left Hand Ditch Co.*,[8] validated the first come, first served priority system of appropriation water rights emerging within mining camps. The 'prior appropriation' system created prioritized rights that: (1) were not tied to land; (2) required beneficial uses, often demonstrated through physical diversions; (3) could be treated as a tradable form of private property subject to protections against impairments to third party interests; and (4) were prioritized to ensure that the first to establish and maintain a beneficial use would be the last to lose access during periods of inadequate supply (Kenney et al., 2009; Tarlock et al., 2002).

Water claims on many Western US rivers outstrip the renewable water available – sometimes even in average years – which has led to overallocation, dewatering and chronic tensions between high-security ('senior') and lower-security ('junior') rights holders. The physical availability of renewable water supplies served as the implicit limit or cap; that is, when rivers run dry (Garrick et al., 2009). This property rights system can be viewed as a natural progression of the law toward economic efficiency: the benefits of defining water rights increased with scarcity, while the costs associated with their definition decreased with technological innovation (Anderson and Hill, 1975). However, principles of distributive justice and equity have been equally important influences on the property rights institutions in this early period (Raymond, 2003; Schorr, 2005; Schorr, 2012). Norms and regulations uphold these notions of equity through the principle of 'no harm' from water transfers, safeguarding third parties against adverse consequences to their water rights from the changes in runoff and return flows caused by water transfers. When considered against the options at the time, the devolved system of private claims imposed intertermporal costs relative to an option with a more proactive and centrally controlled regional or state authority. The resulting water rights system was recognized as paralyzing even at the time, despite their promise of efficiency (Deakin, 1885). The information burden and conflicts created by the maze of water claims and upstream–downstream entanglements has stymied water trade and imposed substantial transaction costs which impede reallocation today.

The second component of institutional development – irrigation infrastructure and associated irrigation supply organizations (mutual companies, districts and associations) – picked up where water rights systems left off by coordinating users to undertake investments in infrastructure and their operations and maintenance (Bretsen and Hill, 2008). The priority-based system was designed to provide security for settlers to encourage investment in irrigation diversion works. Requirements that water rights claimants develop and maintain beneficial uses served as a control against speculation. However, secure water rights were not sufficient to mobilize private capital investment in irrigation infrastructure because economies of scale for such systems were larger than the individual farm. Bretsen and Hill describe the problem of irrigation supply organizations as an issue overcoming the transaction costs associated with reconciling the mismatch between individual farms and irrigation systems: 'farms required a certain scale while irrigation facilities usually required a larger scale . . . [This] led to transaction costs that had to be overcome by farmers and the suppliers of irrigation water' (Bretsen and Hill, 2007: 283).

The institutional solutions to this problem of transaction costs included mutual associations – incorporated and unincorporated – and irrigation districts to reduce the transaction costs associated with vertical integration, controlling opportunistic behavior, and operating and maintaining the canal system. Institutional innovations were increasingly necessary because of the failure of commercial enterprises. The 1894 Carey Act was an initial federal response to the collective action challenge of irrigation infrastructure development which attempted to incentivize private investment. By 1902, the passage of the Newlands Reclamation Act represented a paradigm shift from private to public financing. Large financial investments of almost $22 billion by the federal government addressed challenges of irrigation infrastructure development throughout the Western US; these grants and loans involved long, low-interest payback periods (US GAO, 1996).[9] Irrigation supply organizations were formed as quasi-governmental bodies and granted special powers and eligibility for subsidies and loans. The institutional structures strengthened the position for irrigation districts, with each district able to adopt its own internal decision-making and voting procedures. Although possibly well matched to historical economic imperatives, and available information and monitoring technologies, the prior appropriation water rights and irrigation supply organizations have proven difficult to modernize for contemporary water allocation challenges (Libecap, 2011). Yesterday's innovations have become today's barriers to reform (Eden et al., 2008).

Unlike the Colorado Basin where states adopted variants on the prior appropriation doctrine, states (then territories) of the Columbia Basin embraced the prior appropriation doctrine unevenly. Oregon and Washington opted for a mixture of riparian and prior appropriation rights until eventually encoding the latter as the dominant system for the states' administrative water permits. This reflected, in part, the geographic disparity in rainfall, with wetter conditions in the west and the semi-arid interior tributaries of the eastern parts of the states. This patchwork of property rights systems combined with a diverse set of irrigation water supply organizations to form a heterogeneous institutional geography even today (Chapter 4).

The trajectory of the Colorado had a direct influence in Australia. The dysfunction of the Western US – already apparent in the late nineteenth century – led Deakin to favor state-controlled water licenses. This represented a much weaker property claim than prior appropriation. Like the Western US, however, the era of relative abundance and associated period of water resource development triggered a two-track process of institutional change. The first track was driven by bottom-up changes in norms as miners replaced pastoralists and needed new water rights regimes for irrigated agriculture (Powell, 1989). This pressure overtook pre-existing property claims by pastoralists. This translated into pressure for 'government-sponsored irrigation' to address the collective action challenge of infrastructure financing, operations and maintenance (Harris, 2011).

The second track involved parliamentary action (in other words, more top down) by territorial governments to devise a regulatory scheme to manage infrastructure, irrigation organizations and access to water, shaped by Deakin's experience and vision after visiting the US. The resulting 1886 Victoria Irrigation Act, like *Coffin Ditch*, replaced riparian rights but opted for statutory rather than appropriative rights, controlled by the then colonial government. Davis (1967) noted six major ideas in what has been described as the 'Deakin Water Act': (1) the state eliminated riparian rights and took control over water and vested it in the Crown; (2) the state granted licenses for riparians to divert water for stock and domestic purposes, as well as small gardens, under the common law natural flow principle that diversions by riparians could not impair the natural flows; (3) other types of diversions were prohibited unless a license was granted; (4) a central state licensing agency was needed; (5) 'easements' enabled non-riparians to reach surface water sources; and (6) there were authority and incentives for private irrigators to create local irrigation trusts and receive government loans, building on the 1883 Victoria Water Conservation Act.

The sweeping changes were possible precisely because of its grandfathering of most pre-existing uses to 'continue unaffected by regulation' (Davis, 1967: 653). The Act also promoted local irrigation schemes and initially backed trusts as the institutional apparatus best suited to raise funds and build irrigation systems. The commitment to local administration of irrigation schemes worked initially due to loans, and by 1896, 118 000 acres of irrigation had been created. New South Wales and South Australia adopted similar statutes in 1896 and 1919, respectively. By 1905 irrigation trusts were deemed insolvent and lacked financing to recover costs except during droughts (ibid.: 657). As a consequence, local trusts struggled to stimulate necessary investment or capacity for ongoing operation and maintenance, paralleling the failure of the Carey Act in the US. This capital shortfall required restructuring under the 1905 Victoria Water Act, which intensified land use and required payment for water delivery regardless of use. It is clear that the Murray–Darling diverged from the US in its state-controlled water allocation system (that is, namely that it granted only licenses) and organizational adaptations requiring fees regardless of water use (that is, rather than also requiring beneficial use as in the US).

Despite the contrasting responses to the unsettled problem of irrigation in the US and Australia, the practical and political economic differences are less distinct today. Irrigation communities have assumed increasing political influence over water allocation decisions, even if formal ownership resides with the state. And state governments and irrigators formed a bloc for the purposes of interstate negotiations, even if a division exists between New South Wales (upstream) and South Australia (downstream). Outcomes have converged because irrigation development established powerful vested interests which have pursued common bargaining strategies and political pressure to safeguard the benefits conferred by water access, irrigation infrastructure and interstate apportionment agreements (described below). For example, the transition in the Murray–Darling from centrally administered water licenses to private, tradable water rights in the mid-1980s encountered the strong concerns of irrigators, and corresponding efforts to appease them.

In the state of Victoria – considered the early adopter of state-run irrigation development – path dependency influences contemporary market-based allocations in three ways according to Harris (2011): (1) regulatory protections for irrigators; (2) control of trade into regions with high salinity; and (3) trade restrictions between regions. In sum, despite its conscious divergence from the US, water allocation reform in Victoria has been impeded by the same sources of path dependency: vested interests (irrigation supply organizations) and networked institutions (connected water users, irrigation organizations, and state governments) concerned about knock-on effects of free-flowing water transfers (Harris, 2011; Heinmiller, 2009). Contemporary restrictions on water trading can be viewed as evidence of these vested interests, particularly limits on trade out of irrigation districts and state jurisdictions due to perceived negative impacts.

This mirrored experiences in the Western US. Federal legislation – the 1894 Carey Act – promoted local and private entrepreneurship to build irrigation infrastructure before failure led to new legislation – the 1902 Newlands Reclamation Act – that would restructure irrigation development under increasing federal direction. These decisions would solve the 'unsettled problem' of irrigation in the short term, but establish dependencies and political positions within irrigation communities that have become strongly vested and often inflexible in the face of contemporary pressures. These challenges have become more pressing due to key values and users omitted during the early irrigation development phase, such as groundwater, the environment and indigenous water claims. Responses to the unsettled problem of irrigation left new problems unresolved, but limited flexibility to adapt to these new challenges in an era of limits.

Perceptions of Scarcity: The Hydraulic Era and Interstate Apportionment

> [The Colorado River] apportionment framework is not only based on flawed flow assumptions and ambiguities about how future shortages would be handled, but also contains several notable substantive omissions. Many of these omissions have not been fully addressed as yet, with progress delayed for decades until crises and changes in the paradigm provided a more conducive policymaking environment. (Kenney et al., 2009: 131–132)

The unsettled problem of irrigation pointed to a crisis of capital. Across all three basins, there was the perception that the regions' aridity demanded large-scale infrastructure to coordinate irrigation development and hydropower benefits. In all three rivers, studies and tours were coordinated by central governments (US and UK) to assess the development potential of the rivers. In the Colorado River, for example, river discharge was measured at more than 180 locations between 1889, and the publication of La Rue's classic 1916 report on the *Colorado River and its Utilization* (La Rue, 1916: 10). However, studies of development potential were necessary but insufficient to motivate investment.

Institutional mechanisms were needed to share the benefits and costs generated by scaling up storage infrastructure and distribution systems. This required constitutional agreements about who would bear the major cost outlays involved and how the benefits would be shared. Invoking constitutional authority and case law in the US and Australia, interstate apportionment agreements divided available water among multiple territorial governments (states) within the large basins. Interstate apportionment agreements in the Colorado and Murray–Darling established implicit basin-wide caps by making assumptions about long-term renewable water supplies when determining state apportionments.[10] This spawned the hydraulic mission era and the construction of 'cash register' dams; hydropower revenues were used for multipurpose river basin development with the potential to cross-subsidize irrigation schemes (White, 1957; Worster, 1985; Reisner, 1986). The large water engineering feats of the early twentieth century accelerated irrigation and hydropower development, leading to physical transformations and step shifts in river basin trajectories and their institutional arrangements (Molle and Wester, 2009).

The design of interstate apportionment schemes varied tremendously between two allocation rules: proportional shares and fixed volumes. The former provides a percentage of available flows, while the latter secures an exact volume based on priority (first in time, first in right), location (first in line, first in right), or some other criteria. The Colorado River Compact established a fixed volumetric apportionment between upper- and lower-division states, as well as among the lower-division states of Arizona, California and Nevada. Upper division states later adopted a proportional apportionment among Colorado, New Mexico, Utah and Wyoming. The Columbia never adopted formal interstate apportionment mechanisms despite attempts in the 1960s. The Murray–Darling agreed to a proportional apportionment for the large upstream states and a fixed volumetric delivery for South Australia. Efforts to 'routinize the irregular' in the dam building era required formula and rules for cost and benefit sharing that led to unintended consequences and limited flexibility after flaws in the supply assumptions were exposed (Lach et al., 2005). The distribution of ensuing shortage risks proved uneven and more unfavorable for states and territories slow to develop: that is, the second movers, including both upstream states (in the Colorado River) and downstream states (in South Australia).

Colorado River: fixed volumetric apportionment

The Colorado River Compact of 1922 is notorious for its optimistic assumptions about long-range water supply, which led to chronic overallocation. The seven states in the US portion of the Colorado River divided the basin into two sub-basins and established a ten-year rolling average downstream delivery commitment of 75 million acre feet (equivalent to 92.5 billion m^3) from the upper division to lower division states, along with an additional commitment to Mexico of 1.85 billion m^3 per year under a subsequent 1944 Treaty between the US and Mexico. The ten-year rolling average acknowledged inter-annual variability, but the risks of prolonged drought were underestimated. The residual risk fell on upper basin states to meet fixed downstream delivery requirements. Within the lower basin, California had the first-mover advantage due to private development in the Imperial Valley, and it also wielded political power at the national level at the time of negotiation. As a consequence, California secured 4.4 million acre feet (MAF), leaving 2.8 MAF and 0.3 MAF for Arizona and Nevada, respectively. The upper division states were notionally entitled to an apportionment equivalent to that of the lower division states, but they were slower to develop. The upper division states divided water proportionally among themselves in the Upper Colorado River Compact of 1948. The upper division states faced greater uncertainty about the water available from year to year after meeting downstream delivery requirements. In the Colorado River, fixed interstate apportionments literally locked in the assumptions behind the 1922 Compact and fuelled almost half a century of legal wrangling culminating in the landmark Supreme Court decision *Arizona v. California* (1963). A 1988 Hydrologic Determination recognized limits to the development potential for the upper division states of 6 MAF, which has still not been fulfilled due to financial, physical and political constraints. As recently as the shortage sharing negotiations of 2004–2007, assumptions about Upper Basin development potential have remained a flashpoint for tensions in the Basin.

Murray–Darling: proportional apportionment

Proportional apportionment schemes reduce the risk of climate variability by spreading shortages rather than concentrating the impacts on specific states or users (Fischhendler and Feitelson, 2003). The Murray–Darling is recognized for pioneering such proportional rules in its interstate water apportionment agreement (Turral et al., 2009). The 1914 River Murray Waters Agreement included three elements. First, it planned for a multipurpose and integrated system of water storage and distribution infrastructure with the construction costs shared approximately evenly among the three states of the Southern Murray (Victoria, New South Wales and South Australia). Second, the apportionment rules established a fixed delivery for South Australia (minimum monthly flow, that varied across the year); New South Wales and Victoria received an equal share of the water available at Albury after meeting South Australia's downstream deliveries. The two upstream states retained exclusive control over water from tributaries within their territory. Finally, the Agreement established a small commission to administer the agreement, pointing to the development of one of the first river basin bodies, despite its limited formal authority.

Like the Colorado, the River Murray Waters agreement of 1914–15 established fixed commitments for the most downstream state: 1.85 billion m^3 annually to South Australia. However, South Australia was the downstream state, and unlike California, its downstream counterpart in the Colorado River, it was the third mover to adopt a state irrigation scheme (after Victoria and New South Wales). The upstream states developed first, particularly in Victoria, and adopted a proportional scheme for water sharing. Notably, the agreement included drought provisions covering all three states (Turral et al., 2009). By contrast, the Colorado did not develop shortage rules until 2007, almost a century later, when interests had become entrenched and new demands had developed for cities, indigenous communities and the environment, reflecting intertemporal costs: past institutional design choices deferred drought provisions until the future when the costs of coordination and conflict resolution had escalated with the complexity and diversity of users and interests.

Columbia Basin: no apportionment

The Columbia Basin experiences pronounced spatial variability; semi-arid conditions affect the upstream tributaries rather than the downstream floodplains. As a consequence, the rationale for interstate coordination was comparably limited, initially focused primarily on flood control and fishery benefits, rather than the irrigation diversions where limits proved more relevant within individual upper tributaries than the basin as a whole. Individual states adopted state water laws and administrative permit schemes early in the twentieth century, exemplified by the 1909 Oregon Water Code. The Deschutes River closed to surface water appropriations by the early 1900s (Hubert et al., 2009). State water adjudications (court settlements) defined limits. However, the beneficial use doctrine requires water be put to a beneficial use to maintain water rights; early adjudications rapidly became obsolete in cases where land use change and limited field-based monitoring and administration limited the availability of historical records necessary to validate rights on an essentially continuous basis.

Contemporary efforts to adjudicate and update limits, particularly in the context of mounting impacts of groundwater development on surface water rights, have been constrained, as discussed below in the case of general adjudication of the Yakima River since the 1970s. These agreements centered on tributaries which typically were addressed within states, even on the large systems of the Upper Snake. At the state level, limits on water diversions occurred via adjudications, and in the case of Montana also included compacts, legislation and administrative decisions principally since the early 1990s. The mainstem Columbia is relatively abundant and was not the focus of interstate apportionment until relatively recent disputes between Oregon and Washington. In the rare cases where such tributaries cut across state boundaries, such as the Walla Walla of Northeast Oregon and Southwest Washington, interstate apportionment has been elusive. The 1936 Supreme Court dispute over the Walla Walla, *Washington v. Oregon*, was dismissed. The coordination of hydropower benefits has since dominated multijurisdictional arrangements, even now that new pressures increase the importance of interstate water apportionment schemes (for example National Research Council, 2004).

Comparison: Supply Assumptions, Constitutional-Level Rules, and River Basin Governance

Several aspects of the interstate apportionments combined to reinforce lock-in: constitutional arrangements for interstate apportionment, assumptions about hydrology, first movers, river basin governance arrangements and conflict resolution mechanisms. While water scarcity formed a prime motivation for dam building, hydrological variability was arguably even more important. Flooding, rather than drought or aridity, provided a proximate catalyst for the initial negotiations leading to the 1922 Colorado River Compact and the 1928 Boulder Canyon Act (providing for the construction of the Hoover Dam). The severe flooding of 1905 caused the Alamo Canal to fail and spill into the Imperial Valley, creating the Salton Sea and contributing to the demise of the California Development Company. Irrigators petitioned for a flood control dam and an irrigation canal that would become the All American Canal and deliver water to the Imperial Valley (National Research Council, 2007). Long-term annual average runoff provided a flawed basis for estimating and allocating renewable water supplies. The available historical runoff records did not capture the full range of inter-annual variability and sustained droughts, particularly in the Colorado.

Conversely, the Murray–Darling Basin had experienced a prolonged dry period from 1895 to 1902 (known as the Federation Drought, for its coincidence with constitutional negotiations) that foreshadowed its exposure to inter-annual variability and informed its early water resource development and apportionment decisions (Connell, 2007). Despite the drought of the 1890s in the Western US, flooding was the primary climate extreme in the minds of many negotiators in the Colorado and Columbia (National Research Council, 2007).

The 'flawed flow assumptions and ambiguities about how future shortages would be handled' therefore comprised a common source of path dependent lock-in across the three river basins (Kenney et al., 2009: 131). Efforts to 'drought-proof' agriculture through irrigation paradoxically increased exposure to hydroclimatic risk by increasing irrigated acreage based on average supplies, rather than establishing a buffer for flow variability (following the logic outlined by Garrido, 2011 in eastern Spain *huertas*). States and irrigation communities promoted irrigation based on optimistic assumptions about long-term renewable supplies, particularly in the Colorado, and to a lesser extent in the Murray–Darling where the memory of the Federation Drought faded and gave way to the post-war period of soldier resettlement and efforts to drought-proof irrigation communities. In both cases, the size of irrigation schemes was premised on long-term average availability, with limited room to manoeuvre during prolonged droughts and after the emergence of new users and uses.

The collision of apportionment rules, which were designed to provide certainty, and hydrological variability created the formula for chronic conflict and paralysis. The winners sought to protect their gains from the losers. As a consequence, the distribution of risks within each of the basins has been more skewed toward certain jurisdictions and sectors than originally anticipated, particularly the second movers. This has stoked interstate conflicts and lingering tensions that impede contemporary reforms that would ostensibly benefit (almost) everyone.

The divergent approaches to interstate apportionment and conflict resolution across the three basins influenced the transaction costs of subsequent adaptation efforts (Table 3.3). The fixed allocations for downstream states in the Colorado River, and South Australia in the Murray–Darling, underpin contemporary lock-in and have imposed substantial upfront

Table 3.3 Intertemporal consequences of interstate apportionment: a comparison of apportionment rules, first movers and conflict resolution mechanisms

Basin	Lock-in 'rank'	Apportionment agreement	Allocation rule	First mover	Implementing agency / conflict resolution
Colorado	Highest	Y	Upstream (proportional)	California (downstream)	Bureau of Reclamation
			Upstream–downstream (fixed)		Courts
			Downstream (fixed)		
Columbia	Moderate	N	N/A	N/A	Courts
Murray–Darling	Lowest	Y	Upstream (proportional)	Victoria (upstream)	River Murray Commission
			Upstream–downstream (fixed)		Political negotiation

Note: River basin governance arrangements and conflict management bodies have evolved. The Colorado River has added several multi-stakeholder fora. The Columbia River includes the Northwest Power and Conservation Council and related bodies. The Murray Darling developed the Murray–Darling Basin Commission and later the Murray–Darling Basin Authority.

costs on institutional transitions. In the Colorado River, the second movers in both upstream (state of Colorado) and downstream (Arizona) positions found themselves unexpectedly vulnerable to shortages (National Research Council Committee on Water, 1968). In the Murray–Darling, the second and third movers – New South Wales and South Australia, respectively – lagged behind Victoria. Even the late movers in the Columbia have confronted opposition to the development of the Columbia River Project in Eastern Washington (National Research Council, 2004).

Interstate apportionment schemes were never viewed as permanent fixes, however. The interstate agreements came with river basin governance arrangements to ensure their implementation: a consensus-based river basin commission in the Murray–Darling, and a federal agency serving as a rivermaster in the Colorado. The River Murray Commission was created in 1917 to implement the Murray Waters Agreement in the Murray–Darling, while the Bureau of Reclamation acts as the lead federal agency in the Colorado, formally governed by the Secretary of the Interior as rivermaster, coupled with a strong role of the states and tribes (albeit without the river basin 'Compact' commissions in place as in other Western US rivers). Despite the lack of an apportionment agreement in the Columbia, the Northwest Power and Conservation Council (NPCC) filled a coordination gap; the Bonneville Power Administration (BPA) also wields substantial influence through its funding and planning programs, although the NPCC and BPA lack binding authority regarding interstate apportionment.

The interstate apportionment agreements also depended on conflict resolution mechanisms to interpret the agreements and address emerging disputes. Conflicts and competition dogged negotiations and continued within the bounds of conflict resolution mechanisms established by the interstate agreements. Interstate disputes were a defining feature of river basin development and apportionment even after initial agreements, yet the courts played a bigger role in conflict resolution in the US than in Australia (Challen, 2000). This contrast had profound intertemporal consequences for contemporary efforts to adapt water allocation institutions to new supply and demand patterns. The reliance on costly and contentious legal disputes in the US in both the Colorado (1963, *Arizona v. California*) and the Columbia (1970s, *Washington v. US*) contrasted with experiences in the Murray–Darling where the role of the High Courts has been muted by comparison (Webster and Williams, 2012); but see also Davis, 1967 for a discussion of key court decisions in the 1950s.

High cost decisions and winner-takes-all stakes reinforced lock-in and had a cooling effect on proposals to renegotiate apportionment agreements in light of the new information and values. In the Colorado River, for example, the landmark 1963 US Supreme Court case, *Arizona v.*

California, resolved 40 years of legal disputes between the states to interpret the 1922 compact and confirm interstate apportionment and deliveries to Mexico downstream. The resulting allocation has become encoded in the 1970 long-range operational criteria and adjusted further in light of interim agreements related to coordinated operations for surplus and shortage conditions. Subsequent legislation and administrative criteria constrain water trading and associated reforms in the Colorado River, particularly across state jurisdictions. In the Columbia, indigenous claims to salmon fishing rights in the 1970s triggered two rounds of Supreme Court decisions in the 1974 *US v. Washington* case (Boldt decision), which were then followed by federal listing of 13 species of salmon under the Endangered Species Act in the early 1990s (Blumm, 2002).

Interstate disputes wound through political channels and spirited debates from the very start in the Murray–Darling, occurring in the context of the Convention debates and continuing during negotiation of the River Murray Waters Agreement of 1914/15 (Clark, 1971b, 1971a). This followed bitter disputes during the colonial era when setting the boundaries between the states of Victoria and New South Wales. The Australian High Court interpreted the powers of the River Murray Commission but did not adjudicate the apportionment itself. Political bargaining has continued to dominate within the River Murray Commission and its successors since the 1987 Murray–Darling Basin Agreement: the Murray–Darling Basin Commission (a consensus decision-rule) and the Murray–Darling Basin Authority established by the 2007 Water Act with unilateral authority related to selected provisions of the Act (see Chapter 5 in this volume). As one example of political bargaining within the resulting framework, South Australia adopted internal caps on diversions as early as the 1960s and used its decision in bargaining with upstream states to argue for limiting their development. In both cases, either the courts or political negotiations cemented the distribution of risks and benefits and created perceived winners, losers and vested interests, which hardened antagonistic positions underpinning contemporary politics over caps, allocation decisions and trading rules. But these effects were more pronounced in the US where the courts created a zero-sum calculus that froze allocation institutions until long after there was an impetus for allocation reforms in an era of hard limits and high variability.

Cap, Allocation and Trade: Crises, Hard Limits and Vested Interests

The arrival of 'hard limits' and basin closure occurred when development, demand and drought converged to create chronic imbalances. Early evidence of the arrival of limits can be traced to the mid-twentieth century. The Colorado River experienced the 1950s drought; tree ring studies in the

1970s threw into question the long-term renewable water supplies and raised the spectre of a severe sustained drought long before the unprecedented sequence of dry years since 1999 (Woodhouse et al., 2006; Stockton and Jacoby, 1976; Harding et al., 1995). The Murray–Darling experienced a number of droughts since the Federation Drought of 1895–1902, punctuated by two sustained droughts coinciding with World Wars I and II, and the Millennium Drought from 1997 to 2009 (Helman, 2009). In parallel, major cities grew, and with them came new demands and the emergence of conservation and environmental values. Heightened competition for variable water supplies exposed the intertemporal consequences of past decisions, particularly to address omissions of the previous interstate apportionment schemes (see Kenney, 2009 in the case of the Colorado): the environment, indigenous claims and, to some extent, the water needs of new urban populations.

By the mid-1900s 'the boundaries drawn around responsibilities and authorities were increasingly inadequate for solving emerging problems' (Lach et al., 2005: 2027). New urban and environmental interests clashed with the narrow scope of prevailing water allocation institutions, vested interests of irrigation communities and associated bureaucratic apparatus. It is against this backdrop of new demands that drought and salinity problems triggered an era of hard limits and an increasing need for trade-offs at the basin scale. The ensuing process of capping, allocating and trading water rights has been intensely political and unresolved (Heinmiller, 2006). The recognition of path dependency – that these decisions have long-range consequences – heightens the political struggle to shape the outcomes at each stage. Below, I consider the early stages of cap-and-trade water allocation reforms.

Colorado

In the Colorado River Basin, salinity episodes in the late 1960s and 1970s led to 'Minute 242' (a side agreement) to the international treaty with Mexico; however, the upper basin and state of Arizona continued to develop their water resources. Upstream development had disconnected the river from its delta as early as the 1960s; tree ring studies estimated that historical streamflows were as much as 10 per cent less than the long-range observations of annual runoff (18.5 billion m^2), which were already less than apportionments. However, it was not until 1999 that long-range supply and demand intersected in the Colorado. Long-range demand eclipsed supply by 2002 and Lake Mead reached its historic low (since filling initially) in 2014, with shortages projected for the lower basin as early as 2016. Despite of these looming milestones, the recent Basin Water Supply and Demand Study did not acknowledge the hard limits on supply, and featured several importation schemes among the options for addressing supply-demand imbalances. A group of leading academics in the region - the Colorado River Research Group - released a position paper in late 2014 calling for recognition of limits in the Basin. This announcement was novel in part because it was expressly patterned after parallel efforts at thought leadership and policy advice by scientists in the Wentworth Group of Australia.

Murray–Darling

In the Murray–Darling, the era of limits was recognized as early as the 1960s when South Australia placed a moratorium on new entitlements and reduced existing entitlements based on historic usage (Bjornlund et al., 2013), see Chapter 5 in this volume. New South Wales and Victoria followed suit in the 1980s, ushering a period of property rights reform to unbundle water and land rights. In 1981, the Murray Mouth required dredging for the first time due to the convergence of drought and upstream diversions, a situation repeated again in 2002 for several years during the Millennium Drought. The 1991 algal bloom on the Darling River also punctuated the growing recognition of the basin's limits and water scarcity challenges. A ministerial commission ordered an audit of Murray–Darling water use, which culminated with an interim cap based on water volumes used at the 1993–94 levels of development (Independent Audit Group, 1996).

Columbia

In the Columbia, court adjudication of surface water rights established limits, although this unfolded unevenly within and across states on the US side of the basin (see Chapter 4); some major tributaries closed to new surface water permits in the early 1900s, although groundwater development remained uncapped until much more recently (for example Hubert et al., 2009). The decline of salmon fisheries as early as the 1950s was one sign of the arrival of hard limits. Even though multiple factors contributed to the decline, the dewatering of tributaries for late summer irrigation prevented reproduction (the spawning and rearing phases of the migratory life history). Migratory salmon fisheries use the upper tributaries to spawn and rear before migrating to the oceans, before repeating this life cycle pattern.

Minimum flows and water levels acted as *de jure* limits on appropriable water. Oregon (1955) and Washington (1971) authorized the establishment of minimum flows through administrative rules, and formed watershed and sub-basin jurisdictions to do so, although implementation and enforcement remained weak. This provides an important and relatively early example of the alignment between hydrological and administrative boundaries, as well as the integration of environmental needs into diversion limits. In Idaho (1978) and Montana (1989), instream or minimum stream flow laws played similar roles. The Columbia Basin involved a more piecemeal process in which bottom-up development of diversion limits and minimum flows occurred within a patchwork of state legislative frameworks and, later, basin-wide coordination under the Northwest Power and Conservation Council established in 1980. These fragmented water rights reforms and ad hoc arrangements have lasting consequences, by limiting comprehensive and coordinated reforms needed for landscape scale recovery and for integrated water and watershed planning.

Comparison: politics and path dependency of capping, allocation and trading in three rivers

Processes of capping water use are intensely political, precisely because of vested interests: water was fully allocated or overallocated before limits to sustainable water use had been recognized. Heinmiller (2006) traces the politics of cap-and-trade water policy across three distinct but interlinked phases: capping, allocating and trading water. He argues that each component is shaped by different players and dynamics: 'outcomes of the [capping] process can significantly shape . . . related allocation and trading processes' (ibid.: 455). Thus, the capping becomes a prime source of path dependency, presenting a special challenge when the cap is conceived as permanent rather than temporary and adaptive.

Capping Capping identifies cumulative diversion limits, and in this instance builds on the prior history of interstate apportionment agreements, and the supply assumptions contained therein. Distributive conflicts during the capping phase pit environmental and irrigation interests against one another on the one hand, and upstream versus downstream interests on the other. This configuration of interests and alliances can produce strange bedfellows, such as downstream cities, states and environmental interests working together as a bloc to oppose the development interests of upstream irrigation and cities. However, context and path dependency matter; vested interests depend on the first mover and historic patterns of development and use. In the Colorado River, capping has not occurred at the basin scale; however, the era of limits has pitted new users (cities and environment) against older, vested interests (irrigation); upper basin and lower basin states are therefore not the sole axis of conflict because states have uneven vulnerability. For example, in the lower basin of the Colorado River, California and Arizona have jockeyed to preserve their surplus (California) and limit their shortages (Arizona, Nevada).

Crises and the emergence of new players can transform the politics of capping; for example, limits on water diversions in the tributaries of the Columbia followed the expected pattern of political struggle between fish and farmers until the new residential development and groundwater use in rural areas united environmental interests and senior surface water rights holders. These traditional foes were unified by a common threat: the new pressures from 'amenity migrants' moving from the cities to the urban–rural fringe. Ziemer et al. (2012) write about the transformation of previously antagonistic environmental and farming interests into staunch allies in the face of new groundwater development, leading to court action and efforts to change the Montana Department of Natural Resources and Conservation permitting policy for interpreting basin closure in relation to groundwater pumping. The potential for political realignment demonstrates that path dependency is not synonymous with stasis, but rather a starting point shaped by the past and historic configurations of vested interests.

The Murray–Darling experience, on the other hand, illustrates the steady accretion of political power by vested irrigation interests, despite Deakin's foresight to reserve state control over water resources. The basin-wide audit of water use conducted in June 1995 led to an interim cap based on historical diversions under the infrastructure and rules in place in 1993–94, adjusted from year to year subject to hydroclimatic conditions. Notably, only 61 per cent of authorized water entitlements were actively being used at the time. What would happen to the dormant rights, and what would the intertemporal implications be? Recall that Victoria's Irrigation Law of 1905 had required payment regardless of use, and hence water rights in Australia had long departed from the US rule of 'use it or lose it'. With this precedent, the interim cap acknowledged the legitimacy of unused (sleeper) or underused (dozer) water rights, which set off a race to actively trade them and put them into use, particularly after the cap. These grandfathering decisions made setting the cap politically feasible, but the agreement came with steep (deferred) costs. It led to an increase in water use after the cap and decreased reliability for all entitlement holders who held proportional shares of available water, rather than the fixed volumes provided by the prior appropriation system of the US. The intertemporal consequences of this choice were to defer the costs of establishing sustainable diversion limits until later, to appease the vested interests of historic users and those who held sleepers and dozers. While possibly necessary due to the political economy of property rights reform, this generous grandfathering decision raised the costs of future institutional transitions to sustainable diversion limits by strengthening the claims of irrigators to a private property interest and setting a precedent.

The cap strengthened the political position of private property claims to water by giving economic value to them and incentivizing the activation of unused licenses (Challen, 2000). Subsequent reforms under the 2004

National Water Initiative assigned risk associated with changes in water availability across water users and government, to place the burden of climatic change on users and the burden of policy changes on government (Quiggin, 2011); in many respects, the balance shifted toward a private property view and convergence with the norms in the Western US. While this had benefits in the establishment of water trading, it imposed limits on water management issues taken in the public interest, including the cap, which were considered outside of the market (see Chapter 5). Thus, despite growing consensus about the need for the cap in the mid-1990s, the proverbial devil lies in the detail. This was accomplished in a way that substantially restricted the degrees of freedom to undertake subsequent reforms and anchored the positions of irrigation communities and particularly their lobbyists (see Wheeler et al., 2013, who describe the disconnect between attitudes of irrigators and the actions by their lobbyists). The 2007 Water Act called for a return to sustainable levels of extraction and the establishment of a Basin Plan with new diversion limits. The 2008 amendment to the Act created the Restoring the Balance Program to compensate irrigators for water entitlements converted to environmental uses. It set aside AU$3.1 billion for acquisitions and AU$5.8 billion for irrigation efficiency, placing a price tag on the choices made early in the capping process.

Here the comparison of the intertemporal aspects of capping for river basin trajectories depend on both whether a cap has been established, and if so, how. In the Colorado and Columbia, the cap is implicit, and the Colorado planning efforts even project continued growth in water demand beyond the physical limits with imbalances of almost 4 billion m^3 per year in 2060 (US Bureau of Reclamation, 2012). Despite the shortfalls of its cap, the Murray–Darling has acknowledged limits. It is important not to lose sight of this distinction, which can get lost, particularly for commentators within the Murray–Darling who rightly acknowledge the steep intertemporal costs imposed by these past design choices.

Allocation Allocation processes assign the initial distribution of rights, although historic use patterns and associated property claims have invariably predated the formal establishment of a cap and the subsequent allocation process. The key players in distributive conflicts over allocation processes include the higher- versus lower-security water rights holders, both within and across state jurisdictions (Heinmiller, 2007). For example, those with lower security seek to strengthen their position at the expense of higher-security claimants. In the cap-and-trade period, allocation is better considered as a continuous process of water rights reform (Bruns and Meinzen-Dick, 2005).

In the Colorado and Columbia Basins, the prior appropriation doctrine led to an essentially continuous ranking of individual appropriative rights and collective project-level or district-level entitlements relative to one another, although the spatial positioning and location of the most secure rights can vary. A hierarchy of rights has been used to allocate water between and within states in three partially overlapping steps: laissez-faire appropriative claims within territories during the colonial period and early statehood, followed by interstate apportionment (in Colorado only), and succeeded by an ongoing allocation process using either administrative permit system or general adjudications within states. After a laissez-faire period when appropriators established claims by posting a notice by a point of diversion, an administrative system for permits developed to formalize the process, with general adjudications or negotiated settlements under way in many subbasins (for example, Gila in the Colorado, Yakima in the Columbia).

Adjudications and negotiated settlements have imposed exorbitant costs by any measure to determine the extent, validity and priority of water rights (for example, the *Rettkowski* court case in Washington establishing a case-by-case determination of the validity and extent of historic rights). The Gila River Adjudication of the Colorado (started in 1974) and the Yakima River Adjudication of Washington (started in 1977) have entailed numerous hydrological studies, legal proceedings and negotiations without an apparent end in sight (Feller, 2007). Federal water projects developed distinct systems for water permitting through contracts between irrigation organizations and the Bureau of Reclamation or the relevant state agency. Within irrigation organizations, allocation systems often relied on shares, on a proportional basis, although the decision-making and rules vary by type of irrigation organization. Potential for trading was much higher within than across irrigation district boundaries for a variety of institutional and infrastructure reasons (Thompson Jr, 1993; Ruml, 2005). This created heterogeneity and high information costs for monitoring outside of the districts. The evolution of water allocation institutions has been profoundly shaped by the legal and political economic constraints imposed – often consciously – by past decisions. The intertemporal costs of the prior appropriation system have led to the paralysis predicted by Deakin with pockets of innovation and reform.

Australia's water rights system is based on allocation authority vested in states by Section 100 of the 1901 Constitution, which allowed differences across the states and contributed to lock-in due to the difficulties of moving water outside of state and irrigation jurisdictions. States have adopted different strategies in establishing licenses, with Victoria opting to create higher-security rights (delivered with high reliability), and New South Wales opting for relatively lower-security rights (to support more expansive broad acre crops, but with lower reliability). The creation of higher-security rights in Victoria has fostered permanent plantings in comparison to the annual plantings prevalent in New South Wales, where many entitlements have relatively lower reliability. This has made New South Wales water uses typically lower value, and under pressure during prolonged dry periods, a structural legacy of past decisions that has reinforced opposing positions within basin-wide planning (capping) and allocation efforts.

Trading The politics of trading revolve around the transaction as the unit of analysis. The main players are the buyers, sellers, and third parties other than the buyer and seller who are affected by the trade either directly (downstream rights holders) or indirectly (the irrigation industry) (Heinmiller, 2007). The latter category is expansive and includes the wider irrigation communities and emerging stakeholders. In the Colorado Basin, the main buyers are the cities (for example, Las Vegas, Los Angeles) from agricultural sellers, particularly relatively lower-valued irrigation of alfalfa, negotiating complex transactions to secure temporary or permanent water supplies. In the Columbia, environmental organizations and groundwater-dependent residential developers in the interior basin (for example, Bend, Oregon; Yakima, Washington) seek water from agricultural users. In the Murray–Darling, high-value agriculture (for example, horticulture) buys from lower-value agriculture during drought. The spatial configuration of these buyers and sellers is a function of biophysical characteristics (for example, soil) but also institutional, social and economic.

'Third party effects' has become the term commonly used to capture the effects of trade on stakeholders other than the buyer and seller (Colby, 1990b, 1990a). Historical choices about the regulation of third party impacts therefore underpin path dependent reforms and affect the ability to establish and implement cap-and-trade water allocation reforms. Third party effects have been the source of contrasting regulatory approaches in the Western US and Australian contexts. In the former, a strict no-harm standard has been adopted. Pilz (2006) describes this in detail within Oregon where water transactions for environmental restoration have encountered resistance and irrigators have invoked regulatory safeguards protecting existing rights holders from experiencing harm. Complying with these restrictions involves substantial transaction costs to compile information, verify historic water rights, negotiate price, resolve conflicts, and so on. In Australia, water trading rules have similarly protected against negative social and environmental impacts from trade; however, the interpretation and implementation of these rules are less restrictive because rights are much more standardized as shares of the consumptive pool.

Third party impacts of trade include a suite of technical, political and economic issues, which affect the trajectories and transaction costs of trade, including:

- Area of origin concerns; that is, impacts of trade on communities due to the perceived loss of economic development potential in upstream regions which are net exporters of water (MacDonnell, 2008).
- Return flow issues; that is, impacts of trading on the reliability for upstream and downstream uses (Pilz, 2006).
- Irrigation districts face fixed operations and maintenance (O&M) costs, which are spread across fewer users and create system inefficiencies; for example, the 'Swiss cheese effect' whereby tail-end deliveries get cut off because of insufficient hydraulic head to deliver water by gravity to the end of the canal.

In this context, Grafton et al. (2012) note the need for institutional reforms to improve water trading by addressing the size, duration and distribution of third party impacts. The ensuing enactment of market-enabling water policy reform has occurred from the mid-1980s in each region, albeit in different ways and at different paces. Despite these divergent paths of market-enabling reform and their consequences for transaction costs and adaptive efficiency, Grafton et al. (2012) note a set of common elements: decoupling water and land rights, regulatory oversight and infrastructure. Garrick et al. (2009) operationalize these elements in the context of water markets in overallocated regions where environmental flows are being incorporated into policy design. These steps include:

1. establishment of tradable rights to and cumulative limits on freshwater extraction and alteration;
2. recognition of the environment as a legitimate water user; and
3. authority to transfer existing water rights, including the potential for transfers of high security rights to serve unmet environmental needs.

All three basins have achieved these enabling conditions in different forms, owing to their divergent river basin trajectories and the water rights systems inherited and adapted from the past. A comprehensive solution is impossible at the initial design phase, requiring adaptive mechanisms to address unintended consequences and shifting social and environmental preferences. Piecemeal efforts to establish limits, recognize the environment and promote water trading in the Colorado and Columbia Basins have been under way for at least three decades (National Research Council, 2004; Kenney et al., 2011). The reform process in the

Murray–Darling was piecemeal initially at the state level, but has become more coordinated by a mixture of formal and informal institutions, as discussed in more detail in Chapter 5 in this volume.

In the context of past institutional evolution, water trading activity in the Western US and Australia has expanded, captured by a range of studies toward the end of the first decade of the 2000s (for example, Grafton et al., 2012; Brewer et al., 2008; National Water Commission, 2011). More than $4 billion was spent on water leases and purchases from 1987 to 2008 in 12 of the Western US states (Grafton et al., 2012). Trading in the Murray–Darling involves up to AU$2 billion (2008 dollars) value of water entitlements and allocations traded annually, and almost 40 per cent of surface water use traded in 2011–12 (Grafton and Horne, 2014); see Chapter 5). The Colorado has been the site of increasing activity within states, albeit through complex negotiated transactions between farms and cities (for example, Southern California, Colorado). The spatial pattern of trade varies within the basins: relatively localized in both the Colorado and Columbia, while it expands to include interstate trade and federal buybacks in the Murray–Darling.

THE INITIAL CONDITIONS FOR CONTEMPORARY REFORMS: THE ROLE OF CRISES, PATH DEPENDENCY AND ADAPTIVE EFFICIENCY IN UNLOCKING THE PAST

The conventional wisdom suggests that it is important to never waste a crisis, particularly a sustained drought or a flood (Grafton et al., 2013). The Murray–Darling Basin Authority, for example, has catalogued 23 extreme events as part of its chronology of hydroclimatic risks and responses. Returning to the vantage point from 2004, the institutional evolution of the three basins shifted course during drought crises: the Millennium Drought of the Murray–Darling (1997–2009); the unprecedented sequence of dry years (according to the instrumented record) since 1999 in the Colorado; and the annual droughts of 2001 and 2005 in the Columbia, which magnified chronic challenges and created new coalitions in favor of change.

The forthcoming chapters examine the responses during these policy windows. The Colorado has averted, temporarily at least, a repeat of the winner-takes-all era of conflict resolution that prevailed from the passage of the Compact until the resolution of *Arizona v. California*. Since 2004, it has adopted the interim guidelines for shortage and Minute 319, and it has undertaken a basin-wide study to systematically map out the projected supply and demand imbalances and potential management options to

2060 under climate change scenarios, noting an almost 4 billion m^3 annual average deficit by 2060 under baseline assumptions. This study, along with the impending first lower basin shortage, has triggered a race for solutions, including the consideration of demand-side options such as voluntary water transfers, although these remain a fairly minor component of portfolios. As this book goes to press in 2015, the newest innovation includes a pilot system conservation program involving lower basin municipalities funding conservation to reduce water demand and maintain storage levels to avert shortage triggers (reservoir elevations) in Lake Mead.

A comparative perspective on path dependency and institutional change in the three rivers underscores the need for performance measures and evaluation criteria that account for time and complexity – the defining features of adaptive efficiency. This becomes the focus of the next two chapters about the emergence (Columbia Basin), and the maturation and scaling up of integrated water markets and basin planning (Murray–Darling). The design challenge is therefore to preserve the benefits of institutional stability without imposing unnecessary limits on future institutional flexibility. The foregoing comparison of the three basins across similar stages of development and components of reform illustrates the dynamic tension between stability and flexibility, along with the costly concessions required to overcome the vested interests created by past decisions and technologies. While the costs of lock-in are easiest to recognize in hindsight (for example, through policy lags), current and future alternatives can be assessed in terms of their capacity to preserve future institutional flexibility by limiting tools available to block change (Marshall, 2013).

How can transaction costs analysis and associated concepts of path dependency and adaptive efficiency be used to understand the past and identify design principles likely to prove adaptive in the future? The next chapters examine institutional design, transaction costs and performance. The Columbia illuminates the potential for allocation reform to unlock adaptive efficiency when multilayered water rights reforms and water governance institutions combine local capacity and higher-level coordination and financing; however, the lack of binding interstate coordination institutions exposes lingering limits to state-wide, let alone basin-scale, outcomes. The Murray–Darling is used to examine the potential and limits for scaling up beyond state jurisdictions, using the concepts of local public economies, transaction costs and institutional collective action to address a widening range of economic goods at the intersection of basin governance and water markets. Chapter 6, a synthesis chapter, considers the implications for theory, policy and practice.

NOTES

1. A 'minute' is a decision of the International Boundary Waters Commission charged with administering the binational treaty dividing waters for the Colorado and Rio Grande Rivers between the US and Mexico. Minutes have the character of an amendment, although ratification by Congress is not required.
2. Intergovernmental Agreement on Addressing Water Overallocation and Achieving Environmental Objectives in the Murray–Darling Basin.
3. *C1* and *C1a* (Basurto and Ostrom, 2009).
4. *C2* and *C3* (Basurto and Ostrom, 2009).
5. Contrast with NAP (2007), which categorizes development across four phases: (a) early exploration and initial forays (1860s to 1920); (b) large-scale development (1920 to 1965); (c) relative surplus and shifting priorities (1965 to mid-1980s); and (d) tightening supplies and increasing demands (mid-1980s to present).
6. Britain provided the common law template of riparian water rights (linked to land ownership) which were suited to the climate and water uses of the region.
7. Lichatowitch (2001) identifies waves of development and resource extraction including beaver trapping (fur trade), salmon canneries, forestry, and so on, which accumulated pressures on salmon fisheries before damming and diversions.
8. 6 Colo. 443 (1882).
9. Government Accountability Office (2014). As of 1994 at the end of the primary era of reclamation investment, the federal government had invested $21.8 billion in 133 projects, with irrigation among their purposes. Assistance measures (akin to write-downs) had reduced irrigation payment obligations to only 47 per cent of their initial level, with only 14 of 133 completing repayment obligations.
10. State governments governed allocation among users and districts.

REFERENCES

AGUILERA-KLINK, F., PÉREZ-MORIANA, E. and SÁNCHEZ-GARCÍA, J. 2000. The social construction of scarcity: the case of water in Tenerife (Canary Islands). *Ecological Economics*, 34, 233–45.

ANDERSON, T.L. and HILL, P.J. 1975. The evolution of property rights: a study of the American West. *Journal of Law and Economics*, 18, 163–79.

BASURTO, X. and OSTROM, E. 2009. Beyond the Tragedy of the Commons. *Economia delle fonti di energia e dell'ambiente*, 52, 35−60

BAUMGARTNER, F.R., BREUNIG, C., GREEN-PEDERSEN, C., JONES, B.D., MORTENSEN, P.B., NEYTEMANS, M. and WALGRAVE, S. 2009. Punctuated equilibrium in comparative perspective. *American Journal of Political Science*, 53, 602–19.

BENSON, R.D. 2012. Federal water law and the double whammy: how the Bureau of Reclamation can help the West adapt to drought and climate change. *Ecology Law Quarterly*, 39, 1049.

BJORNLUND, H., WHEELER, S. and ROSSINI, P. 2013. Water markets and their enviornmental, social and economic impact in Australia. In: MAESTU, J. (ed.), *Water Trading and Global Water Scarcity: International Perspectives*. Abingdon: Taylor and Francis.

BLACKBOURNE, D. 2006. *The Conquest of Nature: Water, Landscape and the Making of Modern Germany*. Norton: London.

BLUMM, M.C. 2002. *Sacrificing the Salmon: A Legal and Policy History of the*

Decline of Columbia Basin Salmon. Den Bosch, The Netherlands: BookWorld Publications.
BRETSEN, S.N. and HILL, P.J. 2008. Water markets as a tragedy of the anticommons. *William & Mary Environmental Law and Policy Review*, 33, 723–83.
BREWER, J., KERR, A., GLENNON, R. and LIBECAP, G.D. 2008. Water markets in the West: prices, trading, and contractual forms. *Economic Inquiry*, 46, 91–112.
BRUNS, B.R. and MEINZEN-DICK, R.S. 2001. Water rights and legal pluralism: four contexts for negotiation. *Natural Resources Forum*, 25, 1–10.
BRUNS, B.R. and MEINZEN-DICK, R. 2005. Frameworks for water rights: an overview of institutional options. In: BRUNS, B.R., RINGLER, C. and MEINZEN-DICK, R. (eds), *Water Rights Reform: Lessons for Institutional Design*. Washington, DC: International Food Policy Research Institute.
CHALLEN, R. 2000. *Institutions, Transaction Costs, and Environmental Policy: Institutional Reform for Water Resources*, Cheltenham, UK and Northampton, MA, USA: Edward Elgar Publishing.
CLARK, S. 1971a. The River Murray question: Part I – Colonial days. *Melbourne University Law Review*, 8, 11–40.
CLARK, S.D. 1971b. The River Murray question: Part II – Federation, agreement and future alternatives. *Melbourne University Law Review*, 8, 215–53.
COASE, R.H. 2012. *The Firm, the Market, and the Law*. Chicago, IL: University of Chicago Press.
COLBY, B.G. 1990a. Enhancing instream flow benefits in an era of water marketing. *Water Resources Research*, 26, 1113–20.
COLBY, B.G. 1990b. Transactions costs and efficiency in Western water allocation. *American Journal of Agricultural Economics*, 72, 1184–92.
COLBY, B.G. 2000. Cap-and-trade policy challenges: a tale of three markets. *Land Economics*, 76, 638–58.
COLORADO RIVER RESEARCH GROUP. 2014. The First Step in Repairing the Colorado River's Broken Water Budget: Summary Report. Colorado River Research Group.
CONNELL, D. 2007. *Water Politics in the Murray–Darling Basin*. Annandale, NSW: Federation Press.
CULP, P.W. 2001. *Feasibility of Purchase and Transfer of Water for Instream Flow in the Colorado River Delta, Mexico*, University of Arizona, Udall Center for Studies in Public Policy, Tucson, Arizona.
DAVIS, P.N. 1967. Australian and American water allocation systems compared. *Boston College Law Review*, 9, 647–711.
DEAKIN, A. 1885. *Royal Commission on Water Supply, First Progress Report, Irrigation in Western America*. Melbourne: Government Printer.
EDEN, S., GLENNON, R., KER, A., LIBECAP, G., MEGDAL, S. and SHIPMAN, T. 2008. Agricultural water to municipal use: the legal and institutional context for voluntary transactions in Arizona. *Envirotech Publications*, 58.
FALKENMARK, M. 1989. The massive water scarcity now threatening Africa – why isn't it being addressed? *Ambio*, 18, 112–18.
FELLER, J.M. 2007. The adjudication that ate Arizona water law. *Ariz. L. Rev.*, 49, 405.

FICKLIN, D.L., STEWART, I.T. and MAURER, E.P. 2013. Climate change impacts on streamflow and subbasin-scale hydrology in the Upper Colorado River Basin. *PLoS ONE*, 8, e71297.

FISCHHENDLER, I. and FEITELSON, E. 2003. Spatial adjustment as a mechanism for resolving river basin conflicts: the US–Mexico case. *Political Geography*, 22, 557–83.

FLESSA, K.W., GLENN, E.P., HINOJOSA-HUERTA, O., DE LA PARRA-RENTERÍA, C.A., RAMÍREZ-HERNÁNDEZ, J., SCHMIDT, J.C. and ZAMORA-ARROYO, F.A. 2013. Flooding the Colorado River Delta: A Landscape-Scale Experiment. *Eos, Transactions American Geophysical Union*, 94, 485–6.

GARRICK, D., DE STEFANO, L., FUNG, F., PITTOCK, J., SCHLAGER, E., NEW, M. and CONNELL, D. 2013. Managing hydroclimatic risks in federal rickers: a diagnostic assessment. *Philosophical Transactions of the Royal Society A: Mathematical, Physical and Engineering Sciences*, 371, 1–26.

GARRICK, D., SIEBENTRITT, M.A., AYLWARD, B., BAUER, C. and PURKEY, A. 2009. Water markets and freshwater ecosystem services: policy reform and implementation in the Columbia and Murray–Darling Basins. *Ecological Economics*, 69, 366–79.

GARRIDO, S. 2011. Governing scarcity: water markets, equity and efficiency in pre-1950s eastern Spain. *International Journal of the Commons*, 5, 513–34.

Government Accountability Office (GAO). 2014. Bureau of Reclamation: Availability of Information on Repayment of Water Project Construction Costs Could Be Better Promoted. GAP–14–764.

GLEICK, P.H., HEBERGER, M. and DONNELLY, K. 2014. Zombie water projects. *The World's Water*. Washington, DC: Springer.

GRAFTON, R.Q. and HORNE, J. 2014. Water markets in the Murray–Darling Basin. *Agricultural Water Management*, 145, 61–70.

GRAFTON, R.Q., LIBECAP, G.D., EDWARDS, E.C., O'BRIEN, R.J. and LANDRY, C. 2012. Comparative assessment of water markets: insights from the Murray–Darling Basin of Australia and the Western USA. *Water Policy*, 14, 175–206.

GRAFTON, R.Q., PITTOCK, J., DAVIS, R., WILLIAMS, J., FU, G., WARBURTON, M., UDALL, B., MCKENZIE, R., YU, X. and CHE, N. 2013. Global insights into water resources, climate change and governance. *Nature Climate Change*, 3, 315–21.

GREY, D. and SADOFF, C.W. 2007. Sink or swim? Water security for growth and development. *Water Policy*, 9, 545–71.

HANEMANN, W.M. 2006. The economic conception of water. In: ROGERS, P., LLAMAS, M.R. and CORTINA, L.M. (eds), *Water Crisis: Myth or Reality*.

HARDING, B.L., SANGOYOMI, T.B. and PAYTON, E.A. 1995. Impacts of a severe sustained drought on Colorado River water resources. *Water Resources Bulletin*, 31, 815–24.

HARRIS, E. 2011. The impact of institutional path dependence on water market efficiency in Victoria, Australia. *Water Resources Management*, 25, 4069–80.

HARRISON, K. 2013. Federalism and climate policy innovation: a critical reassessment. *Canadian Public Policy*, 39, S95–108.

HEINMILLER, B.T. 2007. The politics of 'cap and trade' policies. *Natural Resources Journal*, 47 (2), 445–67.

HEINMILLER, B.T. 2009. Path dependency and collective action in common pool governance. *International Journal of the Commons*, 3. Available at: http://www.thecommonsjournal.org/index.php/ijc/article/viewArticle/URN%3ANBN%3ANL%3AUI%3A10-1-100054/49.

HELLER, M.A. 1998. The tragedy of the anticommons: property in the transition from Marx to markets. *Harvard Law Review*, 111, 621–88.

HELMAN, P. 2009. Droughts in the Murray Darling Basin since European settlement. Canberra: Griffith Centre for Coastal Management Research Report No. 100.

HORNE, J. 2013. Australian water policy in a climate change context: some reflections. *International Journal of Water Resources Development*, 29, 137–51.

HUBERT, G., GOLDEN, B. and MCCAULOU, S. 2009. Permanent environmental flow restoration through temporary transactions. International Conference on Implementing Environmental Water Allocations, Port Elizabeth, South Africa.

INDEPENDENT AUDIT GROUP. 1996. *Setting the Cap: Report of the Independent Audit Group*. Canberra: Murray–Darling Basin Commission.

INGRAM, H. and FRASER, L. 2006. Path dependency and adroit innovation: the case of California water. *Punctuated Equilibrium and the Dynamics of US Environmental Policy*, 78–109. Available at: http://www.lhc.ca.gov/studies/activestudies/calfed/IngramSupplementOct05.pdf.

KAY, A. 2005. A critique of the use of path dependency in policy studies. *Public Administration*, 83, 553–71.

KENNEY, D., BATES, S., BENSARD, A. and BERGGREN, J. 2011. The Colorado River and the inevitability of institutional change. *Public Land and Resources Law Review*, 32, 103–52.

KENNEY, D.J. 2009. The Colorado river: what prospect for 'a river no more'. In: MOLLE, F. and WESTER, P. (eds), *River Basin Trajectories: Societies, Environments and Development*. Colombo, Sri Lanka: IWMI.

KRUTILLA, K. and KRAUSE, R. 2010. Transaction costs and environmental policy: an assessment framework and literature review. *International Review of Environmental and Resource Economics*, 4, 261–354.

LA RUE, E.C. 1916. *Colorado River and its Utilization*. Washington, DC: US Government Printing Office.

LACH, D., INGRAM, H. and RAYNER, S. 2005. Maintaining the status quo: how institutional norms and practices create conservative water organizations. *Texas Law Review*, 83, 2027–53.

LIBECAP, G.D. 2011. Institutional path dependence in climate adaptation: Coman's 'Some unsettled problems of irrigation'. *American Economic Review*, 101, 64–80.

LICHATOWICH, J.A. 2001. *Salmon Without Rivers: A History of the Pacific Salmon Crisis*. Washington, DC: Island Press.

MACDONNELL, P.C. 2008. *Protecting Local Economies*. Boulder, CO: The State of Washington.

MARSHALL, G. 2005. *Economics for Collaborative Environmental Management: Regenerating the Commons*, London: Earthscan.

MARSHALL, G., CONNELL, D. and TAYLOR, B. 2013. Australia's Murray–Darling Basin: a century of polycentric experiments in cross-border integration of water resources management. *International Journal of Water Governance*, 1, 197–218.

MARSHALL, G.R. 2013. Transaction costs, collective action and adaptation in managing complex social–ecological systems. *Ecological Economics*, 88, 185–94.
MCCOOL, D. 1987. *Command of the Waters: Iron Triangles, Federal Water Development, and Indian Water*. Tucson, AZ: University of Arizona Press.
MCDERMOTT, T. 1998. Knee-deep disputes for 'water buffaloes'. *Los Angeles Times*, 1 November.
MOLLE, F. and WESTER, P. 2009. River basin trajectories: an inquiry into changing waterscapes. In: MOLLE, F. and WESTER, P. (eds), *River Basin Trajectories: Societies, Environment and Development*. Colombo, Sri Lanka: IWMI.
NATIONAL RESEARCH COUNCIL. 2004. *Managing the Columbia River: Instream Flows, Water Withdrawals and Salmon Survival*. Washington, DC: National Academies Press.
NATIONAL RESEARCH COUNCIL. 2007. *Colorado River Basin Water Management: Evaluating and Adjusting to Hydroclimatic Variability*. Washington, DC: National Academies Press.
NATIONAL RESEARCH COUNCIL COMMITTEE ON WATER. 1968. *Water and Choice in the Colorado Basin: An Example of Alternatives in Water Management: A Report*. Washington, DC: National Academy of Sciences.
NATIONAL WATER COMMISSION (NWC). 2011. *Water Markets in Australia: A Short History*. Canberra: NWC.
NORTH, D.C. 2006. *Understanding the Process of Economic Change*. Princeton, NJ: Academic Foundation.
OSTROM, E. 2011. Reflections on 'some unsettled problems of irrigation'. *The American Economic Review*, 101, 49–63.
OSTROM, E. and BASURTO, X. 2010. Crafting analytical tools to study institutional change. *Journal of Institutional Economics*, 7, 317–43.
OSTROM, E. and COLE, D.H. 2010. *Property in Land and Other Resources*. Cambridge, MA: Lincoln Institute of Land Policy.
PIERSON, P. 2000. Increasing returns, path dependence, and the study of politics. *American Political Science Review*, 94, 251–67.
PILZ, R.D. 2006. At the confluence: Oregon's instream water rights in theory and practice. *Environmental Law Journal*, 36, 1383–420.
PITTOCK, J. 2013. Lessons from adaptation to sustain freshwater environments in the Murray–Darling Basin, Australia. *Wiley Interdisciplinary Reviews: Climate Change*, 4, 429–38.
POWELL, J.M. 1989. *Watering the Garden State: Water, Land, and Community in Victoria, 1834–1988*. Sydney: Allen & Unwin.
QUIGGIN, J. 2011. Managing risk in the Murray–Darling Basin. *Basin Futures*, 47. Available at: http://onlinelibrary.wiley.com/store/10.1029/2010WR009820/asset/wrcr12869.pdf?v=1&t=i58z926e&s=b992a34b8510c4c7fc3ce560675196454ab69988.
RAYMOND, L.S. 2003. *Private Rights in Public Resources: Equity and Property Allocation in Market-Based Environmental Policy*. Washington, DC: Resources for the Future.
REISNER, M. 1986. *Cadillac Desert: The American West and Its Disappearing Water*. New York: Penguin Books.
RUML, C.C. 2005. The Coase theorem and Western US appropriative water rights. *Natural Resources Journal*, 45, 169–210.

SCHLAGER, E. and BLOMQUIST, W.A. 2008. *Embracing Watershed Politics*. Boulder, CO: University Press of Colorado.

SCHORR, D. 2012. *The Colorado Doctrine: Water Rights, Corporations, and Distributive Justice on the American Frontier*. New Haven, CT: Yale University Press.

SCHORR, D.B. 2005. Appropriation as agrarianism: distributive justice in the creation of property rights. *Ecology Law Quarterly*, 32, 3–71.

STOCKTON, C.W. and JACOBY, G.C. 1976. Long-term surface-water supply and streamflow trends in the upper Colorado River Basin based on tree-ring analyses. *Lake Powell Research Project Bulletin*, 18, 1–70.

TARLOCK, A.D., CORBRIDGE, J.M. and GETCHES, D.H. 2002. *Water Resource Management: A Casebook in Law and Public Policy*. New York: Foundations Press.

THOMPSON JR, B.H. 1993. Institutional perspectives on water policy and markets. *California Law Review*, 81, 671–764.

TIETENBERG, T. 2002. The tradeable permits approach to protecting the commons: what have we learned? *Drama of the Commons*. National Academies Press.

TURRAL, H.N., CONNELL, D. and MCKAY, J. 2009. Much ado about the Murray: the drama of restraining water use. In: MOLLE, F. and WESTER, P. (eds), *River Basin Trajectories: Societies, Environments and Development*. Colombo, Sri Lanka: IWMI.

US BUREAU OF RECLAMATION. 2007. *Colorado River Interim Guidelines for Lower Basin Shortages and the Coordinated Operations for Lake Powell and Lake Mead. Record of Decision*. Washington, DC: US Department of the Interior.

US BUREAU OF RECLAMATION. 2012. *Colorado River Basin Water Supply and Demand Study*. Washington, DC: US Department of the Interior.

VOROSMARTY, C.J., DOUGLAS, E.M., GREEN, P.A. and REVENGA, C. 2005. Geospatial indicators of emerging water stress: an application to Africa. *Ambio*, 34, 230–36.

WATERMAN, J. 2010. *Running Dry: A Journey from Source to Sea Down the Colorado River*. Washington, DC: National Geographic Books.

WEBSTER, A.L. and WILLIAMS, J.M. 2012. Can the High Court save the Murray River? *Environmental and Planning Law Journal*, 29, 281–96.

WHITE, G.F. 1957. A perspective of river basin development. *Law and Contemporary Problems*, 22, 157–87.

WILLIAMSON, O.E. 1998. Transaction cost economics: how it works; where it is headed. *De economist*, 146, 23–58.

WOODHOUSE, C.A., GRAY, S.T. and MEKO, D.M. 2006. Updated streamflow reconstructions for the Upper Colorado River Basin. *Water Resources Research*, 42. Available at: http://onlinelibrary.wiley.com/store/10.1029/2005WR004455/asset/wrcr10558.pdf?v=1&t=i58x9n83&s=df30e7e47bf6f5539aa9bc6c3a94e5e0f8317259.

WORSTER, D. 1985. *Rivers of Empire: Water, Aridity, and the Growth of the American West*. New York: Oxford University Press.

ZIEMER, L., BATES, S., CASEY, M. and MONTAGUE, A. 2012. Mitigating for growth: a blueprint for a ground water exchange pilot program in Montana. *Journal of Contemporary Water Research and Education*, 148, 33–43.

4. Emerging water markets in the Columbia Basin: transaction costs and adaptive efficiency in environmental water allocation

> At least on paper it [the Columbia Basin] appears to come close to comprehensive integrated management under a watershed-wide authority, and, in fact, it can boast of many positive accomplishments. . .but the two most sought-after goals – salmon recovery and comprehensive integrated management – have remained terribly elusive. These goals remain elusive not simply because of politics but because of boundedly rational people struggling to achieve collective goals in the face of transaction costs.
> (Schlager and Blomquist, 2008: 89)

> [The] bank balance of acquisition money remain[s] quite healthy, as it [has] turned out to be harder than expected to spend the money. (Neuman, 2004: 439–40)

INTRODUCTION

The outlook for adaptive and sustainable water allocation reform in the Columbia was promising in the early 2000s. Almost five decades of legal reforms, landmark court cases and collaborative watershed programs paved the way for innovative market-based mechanisms to restore water to overstretched tributaries needed for salmon and livelihoods. As a migratory fishery, salmon migrate downstream before returning to their natal grounds upstream to reproduce and complete their life history. As a basin-wide 'water user', salmon fisheries offer a useful barometer of basin health. *New York Times* reporter Timothy Egan (2011) once wrote that the Pacific Northwest was anywhere salmon could get to. That territory has been shrinking due to the cumulative effects of dams, diversions and development. While certain parts of the basin are disproportionately important to conservation and restoration, recovery requires basin-scale coordination.

In recognition of this, the Northwest Power and Conservation Council established its Fish and Wildlife Program in 1982. The listing of 13 salmon species as endangered species gave impetus to basin-wide recovery. By 2002, the Government Accountability Office (GAO) estimated $6 billion in salmon recovery expenditures since 1982 (GAO, 2002). The number increased by an additional $5 billion by 2011 (for a total of $11 billion from 1982 to 2011) with approximately $650 million in annual expenditures in 2011, including $221 million in direct expenditures (NPCC, 2011). Within this stream of funding, a relatively small trickle has focused on market-based reallocation of water to flow-depleted salmon habitat.[1] The Columbia Basin Water Transactions Program (CBWPT) was established in 2002 to coordinate environmental water allocation projects designed and implemented by 'qualified local entities' (QLE) – local partners organized along both state and watershed boundaries. As such, the Columbia is in a unique position to shed light on institutional design and performance at the intersection of water markets and river basin governance.

Residential development expanded in the rural of the interior basin in places such as Bend (Oregon), Kittitas County (Washington) and western Montana. Irrigators then faced competition on two fronts. Environmental interests sought water for depleted habitat, while expanding towns and cities increased groundwater pumping. Suddenly, agricultural and environmental interests became aligned in preserving surface water flows by enforcing diversion limits and requiring mitigation. In short, competition adjusted the calculus and winning coalition needed for institutional change.

In this context, policy reforms in the Columbia Basin enabled water rights reallocation for ecological recovery and mitigation of groundwater pumping impacts on surface water rights. The US states of the Columbia Basin have incorporated environmental flows into the design and implementation of water markets. The market-enabling policy reforms establish limits to freshwater use, tradable water rights, environmental flows as legitimate uses, and public and private financing to restore environmental flows through water transactions. The incorporation of environmental allocations into water markets and associated institutional reforms has not been seamless, however, due to the prevalence of transaction costs – the resources required to define, transfer and manage property rights and to solve collective action problems in natural resource allocation (McCann et al., 2005; Cole, 2002).[2] Public financing has lubricated implementation, but transaction costs have caused a lag. Transaction costs stem from multiple interacting sources tied to resource characteristics and users, property rights institutions, and administrative and infrastructure barriers to trade across user districts and jurisdictions (Bjornlund, 2004; McCann and Easter, 2004; Coggan et al., 2010). Progress varies within states as much as across them, which underscores the importance of local allocation institutions, in addition to the state-wide legal frameworks. Strategic investment in institutional transition costs in the form of multi-stakeholder water planning, water rights reform and water banking institutions has unlocked increasing returns in selected sub-basins. In each state, performance is uneven with success stories juxtaposed with sub-basins locked into continued paralysis.

This chapter uses transaction costs analysis to understand the emergence, evolution and performance of market-based environmental water allocation. It evaluates transaction costs and performance trends across the basin and over time in watershed cases along multiple, interacting dimensions of adaptive efficiency. The next section defines and operationalizes adaptive efficiency as a criterion for performance assessment following the framework introduced in Chapter 2. The chapter then presents the research setting; it also elaborates the original dataset and methodology for transaction costs measurement. Following the adage that what gets measured gets managed, the methodology accounts for and explains transition and transaction costs in environmental water reallocation across three interacting performance variables contributing to adaptive efficiency: water recovery (water rights acquisitions),[3] transaction costs per unit of water recovered, and total program budgets for transaction-related expenditures on policy reform (transition) and implementation. The transaction costs and performance trends (2002–10) and implications are the focus of the next section, which defines different categories of sub-basins based on their trajectories and combinations of performance trends. The chapter then examines the institutional evolution of different groups of sub-basins by focusing on the relationship between water rights reform and performance, and considers the implications for adaptive efficiency. The final section identifies the coordination challenges of scaling up through strategic investment in multilevel policy reform to strengthen enabling conditions and adapt to unintended consequences.

ADAPTIVE EFFICIENCY AND TRANSACTION COSTS

Adaptive Efficiency: Defining Institutional Effectiveness

Institutional effectiveness depends on the policy objectives and criteria for assessment. The Columbia Basin Water Transactions Program (CBWTP) – a public–private partnership serving as an umbrella organization for environmental water acquisition efforts in the basin – established three objectives (Hardner and Gullison, 2007): (1) to develop and test market transactions for acquiring water; (2) to increase water and stream-flow in the tributaries of the Columbia Basin; and (3) to restore habitat for fisheries in stream-flow-limited reaches. A third party evaluation of the CBWTP from 2003 to 2006 documented progress on the first two objectives. The analysis documented a diverse portfolio of temporary and permanent transactional tools and increasing water acquisitions (referred to here as water recovered for the environment); it did not, however, compare spatial and temporal trends across and within the states in the basin.

The third objective, habitat restoration, has proven inconclusive; it takes a longer time to demonstrate progress on ecological outcomes, and this objective is affected by other causally relevant limiting factors. Moreover, habitat restoration will require 'sophisticated coordination with other organizations and government agencies' (Hardner and Gullison, 2007: 2). As such, a definitive link between transactional outcomes and ecological impacts is premature at this stage and scale of implementation. Anecdotal evidence of habitat improvements exists in locations where biologically rich tributaries have become reconnected with the mainstem river. Effort is being devoted to institutionalize best practices in compliance and effectiveness monitoring and to develop evaluation protocols to track and test ecological outcomes.

Transaction costs play an important role in effectiveness criteria. The third party evaluation concluded that transaction costs may be inherently high, noting that: 'there may be no easy way to reduce transaction costs, and it may be necessary for CBWTP (and Bonneville Power Administration, BPA) to accept that achieving instream goals may be more expensive than was previously envisioned, and seek greater financial resources to support the activity of QLEs [qualified local entities]' (Hardner and Gullison, 2007).

Transaction costs remain central to debates over program effectiveness today. In 2010, the Independent Scientific Review Board of the Northwest Power and Conservation Council Fish and Wildlife Program surmised that the Columbia Basin Water Transactions Program needed to measure and manage transaction costs (NPCC, 2010): 'Cost monitoring is needed. Thirty six per cent of the budget is to support QLEs. This is a big investment, and CBWTP should systematically evaluate how to keep acquisition and administration costs as low as possible'.

As a consequence, transaction costs are an integral component of institutional effectiveness; however, it is equally important to consider transaction costs alongside a full range of costs and benefits across multiple spatial and temporal scales. This chapter assesses policy effectiveness in terms of adaptive efficiency. Efficiency refers to the least-cost path to a policy objective based on the (full range of) costs per unit of benefit generated. Neoclassical efficiency is a function of the institutional structure within which costs and benefits are assessed; in other words, the institutional constraints are often taken as given (Bromley, 1982, 1989). Adaptive efficiency, on the other hand, reflects the capacity to solve evolving sets of allocation and coordination dilemmas over the long term in a dynamic context, and in the context of pervasive uncertainty and periodic shocks (North, 1994). Adaptive efficiency is helpful to understand long-term trajectories of institutional economic performance in contexts of entrenched path dependencies, complexity, uncertainty, and feedback between policy reform and implementation (North, 1990).

In the context of market-based environmental water allocation, the evaluative criterion of adaptive efficiency is defined in terms of a relationship between three performance indicators over space and time: water recovery (net increase in flow rates through water acquisitions), average transaction costs (implementation costs per unit of water recovered),[4] and program budgets (total financing available to cover implementation costs required for environmental flow targets and to invest in strategies that reduce transaction costs over time)[5] (see Table 4.1).

Adaptive efficiency is reached in settings where market-based environmental water allocation generates: (1) relatively high water recovery levels; at (2) low (or declining) transaction costs; with (3) sufficient institutional capacity (in terms of program budgets) to cover the transaction costs needed to enact enabling reforms, get to scale, and adapt to unintended consequences. Therefore, adaptive efficiency refers to trends and interactions among variables, and it can only be developed and assessed over the long-term.

Table 4.1 Operationalizing adaptive efficiency in environmental water markets

Variable	Trend
Water recovery	Increasing
Transaction costs per unit of water recovered	Decreasing
Program budgets to cover transaction costs and transition costs	Sufficient to get to scale (achieve gains from trade) and adapt water rights and watershed governance institutions in response to unintended consequences and changing preferences

Transaction and Transition Costs: Barrier to Implementation versus Measure of Adaptive Efficiency?

The prevalence of high transaction costs has been cited as a significant barrier to implementation progress in market-oriented environmental water allocation and water markets more generally (Neuman, 2004; Hardner and Gullison, 2007; Colby, 1990b, 1990a). In the Columbia Basin, the first private instream flow transaction occurred seven years after enabling legislation was enacted in Oregon as part of the 1987 Instream Water Rights Act. Other states in the basin encountered a similar implementation lag, and transactional activity has not increased as quickly as anticipated (see Neuman, 2004). This lag stems in large part from the high cost of collective action to establish and manage institutions that address exclusion, externalities and free-riding challenges in water allocation.

The Coasian definition of transaction costs emphasizes the costs of exchange comprising search, bargaining and monitoring costs incurred during changes in the initial assignment of rights to address externalities. McCann et al. (2005) highlight the importance of transaction costs in environmental policy because property rights are incomplete and evolve for complex and contested natural resources. Cole (2002) traces the problem of high transaction costs explicitly to two linked collective action challenges: the costs of exclusion and coordination. In short, institutions and institutional change are costly, particularly in freshwater resource settings that have been overallocated to agricultural users.

Institutional effectiveness criteria often emphasize least-cost paths to a policy outcome and are premised on transaction costs as a barrier to be minimized. However, transaction costs need to be considered in relation to institutional change and adaptive capacity, following the concepts and evidence elaborated in Chapter 2. This becomes particularly important as transaction costs minimization in one period may impose intertemporal costs that contribute to lock-in in subsequent periods, as illustrated by all three basins in the previous chapter. Following Chapter 2 in this volume, transaction costs are defined as the resources required to define, transfer and manage property rights (McCann et al., 2005). Transition costs account for the costs of moving from the institutional status quo to a new arrangement (Challen, 2000; Ostrom, 1990; Basurto and Ostrom, 2009). Marshall (2013) further expands the definition of transaction costs to account for institutional change in complex social-ecological systems.[6]

Therefore, this chapter acknowledges that transaction and transition costs also occur at multiple levels (Williamson, 1998) of policy design and implementation, including the costs of: (1) exchanging property rights within the prevailing institutional setting (that is, static transaction costs); (2) changing the institutional framework governing those exchanges (that is, institutional transition costs); and (3) lock-in, in terms of the lost institutional flexibility imposed by past institutional commitments (North, 1990; Marshall, 2005). These distinctions have antecedents in the sports analogy developed in new institutional economics literature, which separates 'playing the game' from changing 'the rules of the game' (North, 1990; Williamson, 1998). In addition, implementing organizations are the 'teams in the game' which are structured in part to economize on transaction costs through a 'discriminating alignment' between organizational characteristics and characteristics of the resources and economic goods being transacted (Williamson, 1998; Coggan et al., 2013).

A less often noted antecedent to this boundary issue appears in the common pool resource literature that separates collective choice (decision-making) from operational-level rules and actions (Ostrom, 2009; Schlager and Blomquist, 2008). The interaction between levels of action and transaction costs is integral to the evolution and performance of water allocation institutions over space and time. The friction between individual and collective choice underpins the calculus of institutional change elaborated in Chapter 2 based on Basurto and Ostrom (2009).

Public Transaction Costs in Market-Based Public Goods Provision

In assessing transaction costs, an important question is which transaction costs (the types, actors, and so on) are included in the analysis. Most empirical work is partial in nature, in that studies 'have focused on some subset of . . . costs, either regarding the type of cost or who bears them' (McCann et al., 2005: 533). In markets for private goods, the transaction costs of private actors – buyers, sellers and third parties – are of prime interest. In environmental markets for public goods (that is, the use of market mechanisms to preserve or enhance public goods), transaction costs analyses often focus on public sector transaction costs (McCann and Easter, 2000; Mettepenningen et al., 2011).

Mettepenningen et al. (2011) explain the rationale for a focus on public sector transaction costs in the analysis of environmental policy effectiveness: (1) environmental schemes may face declining budgets to reach fixed or intensifying policy challenges; (2) accountability measures require 'value for money' in terms of policy outputs per dollar expended in the program budget; and (3) evidence is needed to inform trade-offs when choosing between competing priorities for public investment. The public and non-profit organizations brokering transactions in the CBWTP confront an explicit motive to economize on transaction costs as the program faces periodic reauthorization.[7]

The focus on public transaction costs is consistent with the objective of evaluating institutional performance and governance arrangements on the demand side of the public goods market for water. This analysis therefore excludes private transaction costs faced by sellers. This decision is accompanied by two simplifying assumptions that should be tested in future empirical research in a comprehensive benefit–cost framework. First, the magnitude of transaction costs incurred by sellers in a public good market is expected to be positively correlated with those incurred by public actors; information and coordination costs are influenced by similar politics, institutional arrangements and local contexts. Second, non-profit conservation brokers and regulatory officials serve as extension officers to incentivize and facilitate participation by sellers, building trust and capacity by underwriting the costs of information and coordination faced by prospective sellers. Extension officers argue that public transaction costs incurred to implement water transactions partially serve to shift the burden from sellers to buyers and regulators (Lovrich et al., 2004; Neuman, 2004). A full accounting of public and private transaction costs and benefits is a promising area of research based on the conceptual and methodological advances developed here for public sector transaction costs.

Transaction Costs and Uneven Performance: The Role of Local Institutions for Collective Action in Water Rights Reform

This chapter investigates the relationship between transaction costs and institutional performance by examining the spatial and temporal patterns of performance in market-based environmental water allocation. These performance trends are decomposed into three interacting performance variables: water recovery levels, transaction costs per unit of water, and program budgets. How do these three performance indicators interact to shape prospects for adaptive efficiency? The hypothesis is that performance will vary as much within states (where authority for water allocation is vested in the Western US) as across them. This expectation is formed based on: (1) the lock-in costs (path dependency) of historic water users, uses and biophysical characteristics of the system;[8] and (2) localized drivers of water rights reform and administration capacity, which reflects the importance of local time- and place-specific information and values (Hayek, 1945).

Previous research establishes the basis for this working hypothesis about intrastate variability in spatial and temporal performance trends.

Colby (1990b) examines the relationship between transaction costs and efficiency in water allocation institutions. Her empirical analysis of 'policy-induced' transaction costs (associated with administrative procedures) reveals that transaction costs are 'higher in areas where water is more scarce and valuable, transfers are more controversial, and the externalities of water transfers are more likely to be significant' (p. 1189). MacDonnell (1990) finds high degrees of intrastate variation in transaction costs due to differences in institutional capacity to respond to scarcity, conflict and externalities.

Ruml (2005) provides the basis for the hypothesis in his work on the Coase theorem in the Western US system of prior appropriation. His analysis suggests two different institutional contexts for water transactions in the Western US: (1) an appropriative system under statutory rules; and (2) irrigation district-level systems governed by water user rules. The former is prevalent at the state level, where water transactions can become trapped in a vicious cycle of ill-defined and insecure property rights with high transaction costs; the latter (district level) has fostered isolated pockets of favorable water market conditions governed by a 'virtuous cycle' whereby water rights are defined clearly and transferred with low transaction costs. Carey et al. (2002) describe an example of the virtuous cycle via the emergence of an enabling framework within informal networks of 'affiliated farms' in the Westlands Water District in California. In that case study, fixed transaction costs are low enough to enable trading for transactions of relatively small volumes, the bellwether of a maturing market.

In the context of market-based environmental water allocation, state-level water rights reforms are a necessary condition. However, adaptive efficiency is expected to depend on additional reform and institutional capacity at the sub-basin scale to access local time- and place-specific access to information and coordination (Hayek, 1945). This chapter documents evidence for this thesis by elaborating and testing the transaction costs evaluation framework to assess spatial and temporal performance trends within and across state jurisdictions.

The Columbia Basin: A Laboratory of Market-Based and Place-Based Initiatives

Water allocation in the Columbia Basin is governed under the Western US prior appropriation doctrine, which stipulates that the first to establish and maintain a beneficial use is the last to lose access during periods of inadequate supplies. Ecosystems tend to receive the residual flows. The allocation system allows a chronic imbalance between legal rights and claims (that is, 'paper water') and physical availability (that is, 'wet water').

Implementing organizations in the Columbia include non-profit water trusts, watershed organizations, indigenous communities, and public agencies that operate at multiple scales. The Northwest Power and Conservation Council (NPCC) has coordinated local recovery efforts along eco-regional boundaries by defining sub-basins: zones of ecological interactions. The Columbia Basin is particularly well suited to test this hypothesis because four decades of watershed governance reforms ensure that field-level water rights administration and watershed planning align closely with eco-regional boundaries (NPCC, 2005). Sub-basins are therefore the appropriate scale at which to analyze efforts to address the externalities of overallocation incurred by ecosystems and existing water users.

The Columbia Basin includes 62 sub-basins. Twelve sub-basins and one eco-province ($n = 13$) comprise the study area in the first period of analysis (2003–07) and were chosen based on the need for and presence of transactional activity to address low-flow limitations (Table 4.2). Of these, ten remained[9] active within the program during the second period of analysis (2008–10) within the CBWTP, although new sub-basins were being incorporated in the second phase and should be the focus of future research. Sub-basins were selected in the study design because of significant unmet demand, which led to their inclusion as priority sub-basins in the CBWTP. The Upper Snake water banking pool in Idaho is the only Columbia sub-basin with substantial transactional activity that is excluded; legal restrictions in Idaho and the dominance of the federal water project crowd out local and decentralized market-based environmental water allocation.

During the initial period of implementation, no systematic mechanisms were available to prioritize transaction projects based on the marginal ecological benefit or the proportion of unmet demand addressed through the transaction. Sub-basins were selected for their common need to increase flow-dependent habitat for salmon. High levels of unmet demand serve as a control for case selection, but the relative need (and hence marginal ecological and economic benefits of transactions) is assumed to be similar across sub-basins, particularly during the early stages of implementation.

These cases enable comparisons both within and across state boundaries in areas where basic enabling conditions and initial impediments to environmental water transactions had been addressed prior to the formation of the CBWTP in 2002. The cases encompass sub-basins in Idaho, Montana, Oregon and Washington, the primary riparian states on the US side of the border.

A series of contextual factors is noteworthy (Table 4.2). These factors are not expected to have a determinative impact on the direction or magnitude of any of the three performance variables because high or low levels of performance are expected to be possible regardless of the

Table 4.2 Case study sub-basins and contextual factors

	Sub-basin	Physical			Institutional			Socio-economic		
		Size (km^2)	Federal storage	Anadromous	Adjudication	Groundwater mitigation	Water banking	Seller organization	Urban character	Environmental water prices (USD)
Idaho	Salmon	36 262	No	**Yes**	Yes	No	Yes	Dispersed	Rural	$2394
Montana	Bitterroot	7399	No	No	Partial	No	No	Dispersed	Mixed	$1445
	Blackfoot	5997	No	No	Partial	No	No	Dispersed	Rural	$5844
	Clark Fork	20 770	No	No	Partial	No	No	Dispersed	Rural	$3107
	Flathead	21 934	No	No	No	No	No	Dispersed	Rural	$8170
Oregon	Deschutes	27 744	Yes	**No***	Yes	Yes	Yes	District	Mixed	$5883
	Grande Ronde	10 613	No	Yes	Yes	No	No	Both	Rural	$8808
	John Day	20 527	No	Yes	Yes	No	No	Dispersed	Rural	$2848
	Umatilla	6607	Yes	Yes	Yes	No	No	District	Rural	$1202
	Willamette	29 050	No	Yes	Yes	No	No	Both	Urban	$2516
Washington	Upper Columbia	–	No	Yes	Partial	No	No	Both	Rural	$5715
	Walla Walla	4558	No	Yes	Partial	Yes*	Yes	Both	Mixed	$3982
	Yakima	15 990	Yes	Yes	Partial	Yes*	Yes	Both	Mixed	$9416

Notes:
* Denotes that the characteristic changed during the early implementation period from 2003–2007;
Environmental water price data for 2003–2007 source is described in Garrick and Aylward 2012.
Bolded sub-basins ('Anadromous' column) had established a sub-basin set of flow targets through planning processes that guided implementation efforts during the period of analysis.

contextual feature. The physical, institutional and socio-economic characteristics of the sub-basins can vary as much within the states as across them and therefore favor a mixture of local and state institutional capacity well matched to the local context. The development and strengthening of enabling conditions has required institutional transitions to reform water rights and river basin institutions. These multi-level institutional transitions establish a strong fit between local conditions and institutional responses needed for high levels of performance and adaptive efficiency as outlined above.

The key physical characteristics include the presence or absence of reservoir storage in the form of a federal storage project and associated infrastructure, an environmental flow target (or set of targets), and anadromous (migratory) or resident fisheries. Although a range of studies identify storage as a means of establishing homogenous rights and lowering monitoring costs (for example Garrido, 2011), the projects equally bring in higher-level institutional actors (particularly federal agencies) whose involvement may incur high transaction costs. The institutional characteristics include the presence or absence of a water rights adjudication (a court decision to define water rights), a groundwater mitigation rule, and/or a water banking structure. The presence of all three characteristics may be associated with well-defined rights involving lower transaction costs, but also the potential for higher competition and consumptive use values of water, which may increase conflict associated with reallocation proposals. Socio-economic characteristics are defined by the presence of irrigation districts (and associated infrastructure) versus dispersed users, the proximity to an urban center, and competition for water (as measured via average price per unit of environmental water acquired through the CBWTP). Again, the presence of all three components is typically linked with low transaction costs, but increase the transaction costs associated with district participation and potential conflicts between urban and environmental users.

The cases are all defined by the presence of transactional activity and coordinated by the CBWTP. With regard to water institutions, biophysical setting and socio-economic characteristics in the Columbia Basin, the following simplifying assumptions apply to the cases selected:

- Legal and institutional reforms all occur within the broader framework for water allocation under the prior appropriation doctrine, and the initial enabling conditions for water trading have been met, as specified in Chapter 3.
- All sub-basins have high levels of unmet ecological demand where instream flow has been identified as a limiting factor for fish

recovery. Insufficient data and analysis exist to compare transactions in terms of their marginal ecological benefits.[10]

- Sellers face roughly similar types and magnitudes of transaction costs across the sub-basins; the role of conservation brokers is to serve as extension agents conducting outreach with potential willing sellers to decrease the information costs of learning about new incentives and reallocation options.

WHAT GETS MEASURED GETS MANAGED: DATA AND METHODOLOGY FOR TRANSACTION COSTS ACCOUNTING

In the early 1990s environmental economists began to measure transaction costs to examine institutional barriers and performance trends in water markets across the Western US, Chile and Australia (Howitt, 1994; Hearne and Easter, 1995; Easter et al., 1998; Challen, 2000; Colby, 1990a, 1990b; Allen Consulting Group, 2006). This research relied on a range of conceptual frameworks and methodological approaches, limiting the comparability of findings and insights into institutional performance. These studies measured different types of costs at different stages of the policy cycle and across varying subsets of market actors: buyers, sellers, administrators and third parties (McCann et al., 2005). Empirical estimates of transaction costs have yet to examine the use of water transactions to restore environmental flows and related public goods, although Colby (1990a: 1118) cites high transaction costs as a constraint on water marketing activity to enhance instream flows.

This methodology fills this gap in the literature to examine the relationship between 'public' transaction costs and institutional performance (see Mettepenningen et al., 2011). Public transaction costs refer to the costs incurred by a subset of the actors; principally government agencies and non-profits. In the context of overallocated freshwater ecosystems, public transaction costs include policy reform and implementation expenditures by governmental and non-profit actors to reallocate private water use rights into the public trust (Challen, 2000).

The evaluation of policy design and performance compiles data on three interacting performance indicators – water recovery, transaction costs and institutional capacity (the latter simplified here as 'program budgets' to capture a subset of institutional capacity) – that together contribute to adaptive efficiency (Tables 4.1 and 4.3). The transaction is the common unit of analysis (Commons, 1931; and see Chapter 2 in this volume). The sub-basin is the spatial scale. The temporal scale for aggregation is

Table 4.3 Definition of performance variables and explanatory factors

Variable	Description (unit)	Metrics
Water recovery	Net increase in instream flow, discounted into a net present value over the term of the contract (cubic feet per second, CFS)	• Total (by period) • Annual • Annual average
Transaction costs	Financial expenditure on design and implementation of water acquisition projects, including both transition and transaction costs ($/CFS), excluding the price of the water itself	• Average $/CFS (by study period) • Average Annual $/CFS (by year)
Program budgets to cover transaction costs and transition costs	Total financial expenditure on transition and transaction costs ($)	• Total (by period) • Annual • Annual average
Explanatory factors	Sub-basin and state-level coding of water rights reforms	• Exclusion • Transferability • Administration

important. I calculate the transaction costs per unit of water recovered aggregated over two sub-periods during an eight-year period of analysis. The time frame spans two sub-phases: (1) the five-year period of early implementation from the program's first year (2003) until 2007, the completion of the third-party assessment (Hardner and Gullison, 2007); and (2) the subsequent three-year period of restructuring and transition from 2008 to 2010. The third-party evaluation marked the divide between the two sub-periods. The two phases are also distinguished by changes in major exogenous variables, namely the global financial crisis of 2008–09, which combined with the third party evaluation's recommendations to prompt changes in program design, including increased focus on 'value for money' in terms of ecological and cost-effectiveness monitoring.

To examine the temporal trends in transaction costs, I also calculate the average annual transaction costs per unit of water recovered. A simple example can explain the difference between these measures: average versus average annual transaction costs. If five cubic feet per second of water are recovered from 2003 to 2007 for a total transaction cost expenditure of 10 dollars, the average transaction costs are $2/cfs for the sub-basin during this period. However, this masks the variability over time. The $2/cfs could be based on five years with one CFS per year at $2/cfs, or a trend that is increasing or decreasing. If transaction costs are increasing, the first year could include all CFS, followed by four years without water recovery, or some related permutations. If transaction costs are decreasing, the first year could include one CFS at $6/cfs, followed by four years with one CFS per year at $1/cfs, or some related permutation. The transaction costs accounting methodology compiles the transaction costs per unit of water acquired during the full period of implementation and the annual average transaction costs to disentangle spatial and temporal patterns of institutional change and adaptive efficiency.

Water recovery is defined as the increased rate of instream flow (cubic feet per second, CFS) generated through transactional activity to acquire or lease water rights.[11] Notably, transactions include both temporary and permanent deals. To create comparable measures of transactional activity, long-term deals were treated as a form of annuity, and the water recovered over the full term of the deal was discounted into a net present value for the initial year of the transaction, as outlined by Garrick and Aylward (2012). Transaction costs were assessed using financial data and questionnaires (*n* = 58, over the two periods) about expenditures related to transaction design and implementation activity. Program budgets include financial expenditures for transaction design and implementation activity by two entities: non-profit conservation 'brokers' and regulatory agencies. The program budgets expressly exclude the cost of water and expenditures for unrelated program activities. The original dataset on transactions, transaction costs and program budgets was scrutinized to limit error and ensure validity. A verification process involved expert consultation with practitioners to review data sources and collection techniques, including confirmation of survey responses used to account and allocate transaction costs and program budgets. Documentation of metadata (that is, data about the data) enables replication and longitudinal analysis. Garrick and Aylward (2012) provide flowcharts identifying each step in the accounting process, including a detailed summary of the data collection and analysis.

EMERGING MARKETS AND ADAPTIVE EFFICIENCY: TRENDS AND EXPLANATORY FACTORS

The results confirm the expectation that the performance of market-based environmental water acquisition programs will vary as much within states as across them. The maps in Figures 4.1a–c account and aggregate results for the three performance variables over the first study period to depict the intra- and interstate variability for water recovery (net present value of CFS by sub-basin), transaction costs per unit of water recovered, and program budgets of transaction-related expenditures.

Spatial Trends and Uneven Performance: Intrastate Variability Across Both Study Periods: 2003–2007 (Period 1) and 2008–2010 (Period 2)

Water recovery

Sub-basin water recovery levels range from 19 to 1618 CFS (in present value terms) for the first study period (Figure 4.1a) with substantial intrastate variation, and from 12 to 1364 CFS in the second study period. Performance trends confirm the expectation that performance varies as much within states as across them. In the first study period, sub-basins in three of the four states appear in the upper two quantiles for water recovered (Deschutes, Oregon; Yakima, Washington; and Bitterroot and Blackfoot, Montana); each of these three states also has a case in the lower quantiles (Grande Ronde, Oregon; Upper Columbia, Washington; and Flathead and Clark Fork, Montana). This intrastate variation continues in the second period, but with noteworthy differences in the top and bottom performers in each state. For example, the Deschutes remained in the top quantiles but was joined by the Salmon (ID) and Upper Columbia (WA).

Oregon's sub-basins demonstrate the full range of recovery levels from relatively higher water recovery levels in the Deschutes to lower levels in the Umatilla and Grande Ronde, where political resistance in the latter has engendered administrative protests and threats of court challenges (described by Pilz, 2006). Washington reached high water recovery levels in the Yakima, reflecting a pulse of transactional activity for drought response in 2005. Recovery levels were relatively low in the Upper Columbia during the first study- period despite high ecological priorities associated with salmon fisheries in the eco-province's Methow and Okanogan tributaries. The Upper Columbia confronted institutional barriers associated with poorly defined water rights and, until more recently, local resistance to environmental water allocation. The Upper Columbia overcame many of these barriers to become a relatively high performer in the second period.

Idaho's only case, the Salmon, achieved the local water recovery target of 35 CFS of the Lemhi tributary of the Salmon sub-basin. The state government-run Idaho Water Transactions Program focuses on the Salmon, prioritized by state law, and relied on temporary contracts to build toward longer-term agreements. By the start of period 2, the Salmon

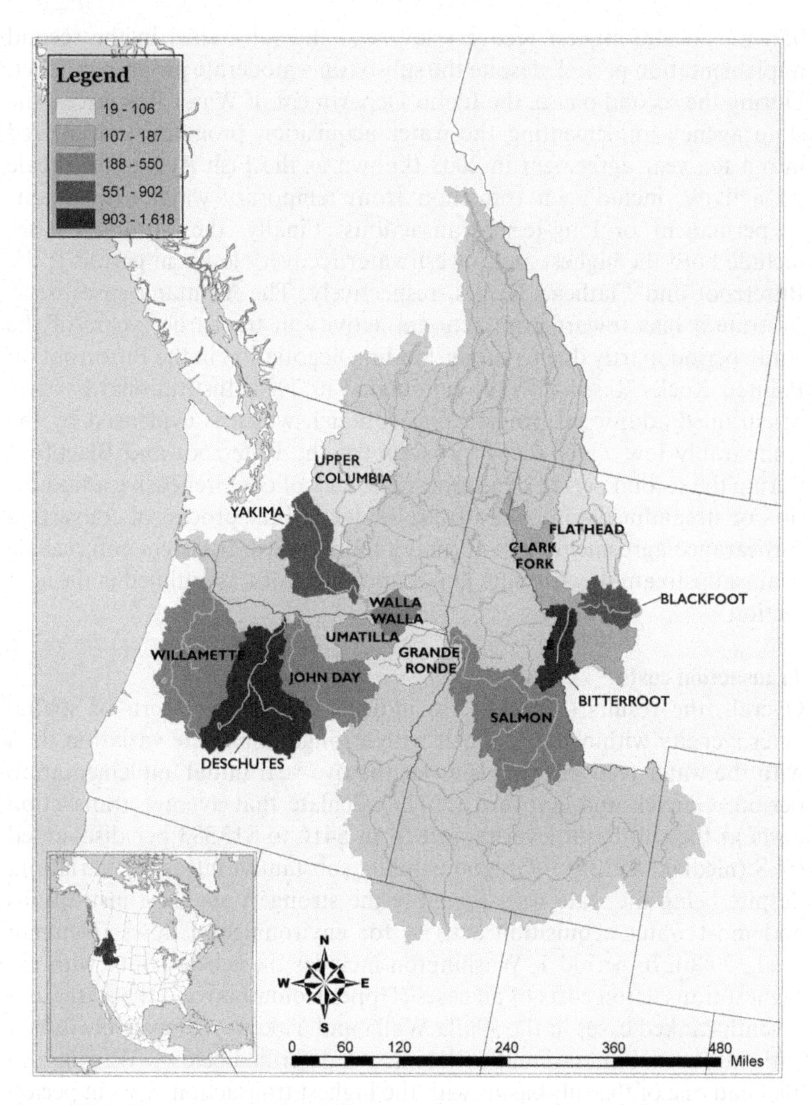

Note: This figure represents the cubic feet per second restored over the first implementation period from 2003–2007. The water recovered by each transaction is treated like an annuity and discounted into a net present value to compare temporary (leases) and permanent (acquisition) transactions using an apples-to-apples metric. The total water recovered (in cubic feet per second) is aggregated at the sub-basin scale for the five-year period. The sub-basins were ranked and mapped using quantiles, as summarized in Garrick and Aylward 2012. Each quantile represents a 20th percentile interval, with darker shades indicating higher levels of each performance variable.

Figure 4.1a Spatial variability in water recovery (net present value of CFS restored), 2003–2007

had the second-highest recovery levels of the sub-basins in the second implementation period, despite the sub-basin's moderate program budget. During the second phase, the Idaho Department of Water Resources, the state agency implementing the water acquisition program had entered into a ten-year agreement in 2008 (known as the Fish Accords) to scale up activity, including a transition from temporary water acquisitions to permanent or long-term transactions. Finally, the Montana cases include both the highest and lowest water recovery levels in period 1: the Bitterroot and Flathead Rivers, respectively. The Montana cases demonstrate a bias toward transactional activity in the earlier years of the study period, partly due to a longstanding negotiation in the Bitterroot on Painted Rocks Reservoir that culminated in 2004. Institutional barriers constrained additional progress in Montana, which is evidenced by the comparably low water recovery levels for the Bitterroot and Blackfoot during the second period of analysis. The lack of comprehensive adjudication or streamlined administrative rules delayed the process of converting forbearance agreements into formal administrative transfers enforceable against upstream/downstream junior appropriators, as outlined in the next section.

Transaction costs

Overall, the results (Figure 4.1b) indicate a similar pattern of spatial heterogeneity within states, albeit with stronger interstate variation than with the water recovery trends. Over the five-year initial implementation period, Garrick and Aylward (2012) calculate that average transaction costs at the sub-basin level ranged from \$416 to \$13 383 per discounted CFS (median: \$2225).[12] Oregon exhibits substantial intrastate variation, despite being the state with arguably the strongest enabling institutions and most water acquisition activity for environmental flows (Neuman et al., 2006). In period 1, Washington includes the eco-province with the highest transaction costs of all cases (Upper Columbia) and the sixth- and seventh-ranked cases in the Walla Walla and Yakima, respectively (albeit with substantial interannual variability, as described below). Washington also had one of the sub-basins with the highest transaction costs in period 2, but in the Walla Walla, potentially a consequence of fixed costs associated with setting up a groundwater mitigation bank during that period. The Upper Columbia reduced its average transaction costs by period 2, while the Walla Walla increased. Finally, the relatively low transaction costs in Montana and Idaho stem in part from dependence on a single actor for implementation, the private non-profits and the state, respectively. Montana lacks agency contributions because initial transactions proceeded outside of the formal administrative procedures through

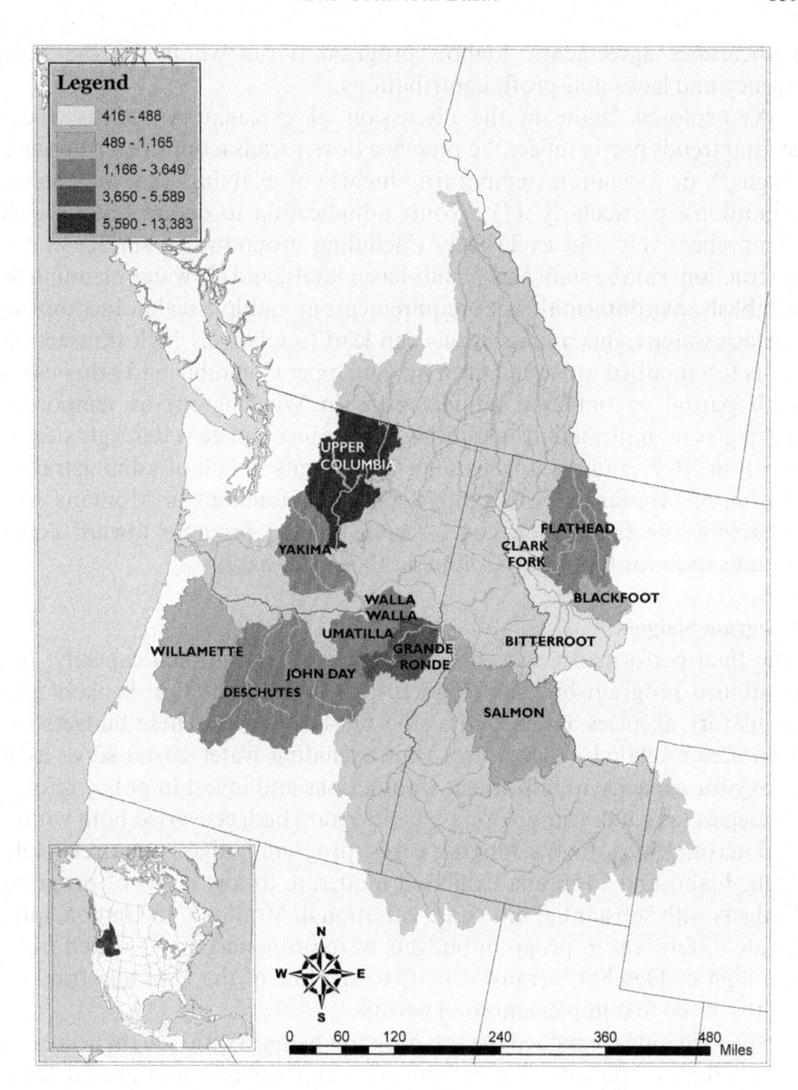

Note: This figure represents the transaction costs per CFS recovered over the first implementation period from 2003–2007. The water recovered by each transaction is treated like an annuity, discounted into a net present value and aggregated at the sub-basin scale for the five-year period. The transaction-related expenditures at the sub-basin level are aggregated over the same five-year period. The transaction-related expenditures are divided by the water recovered to calculate the transaction costs per CFS. The sub-basins were ranked and mapped using quantiles, as summarized in Garrick and Aylward 2012. Each quantile represents a 20th percentile interval, with darker shades indicating higher levels of each performance variable.

Figure 4.1b Spatial variability in transaction costs ($/CFS), 2003–2007

forbearance agreements. Idaho's program is run wholly by the state agency and lacks non-profit contributions.

As explored below in the discussion of explanatory factors, these spatial trends partly reflect the presence (lower transaction costs), absence (higher), or formation (temporarily higher) of enabling legal and policy conditions, particularly: (1) a court adjudication to define water rights comprehensively and exclusively (including groundwater–surface water interactions) at the state and/or sub-basin levels; and (2) water planning to establish environmental water requirements at multiple scales. Incomplete surface water rights adjudications can lead to relatively high transaction costs (exemplified in Washington by the Upper Columbia and other cases with partial or outdated adjudications in Washington) or temporary strategies to implement transactions despite incomplete water rights adjudications (for example, forbearance agreements in lieu of administrative leasing in Montana). For example, initial transactions in Montana had relatively low transaction costs despite limited progress toward court adjudication for the state's Columbia Basin sub-basins.

Program budgets

The final performance variable is a subset of institutional capacity: the combined program budgets (Figure 4.1c) for conservation brokers and regulatory agencies aggregated at the sub-basin level. These budgets for transaction-related expenditures (but excluding water costs) serve as a proxy for capacity to implement transactions and invest in policy reform strategies to reduce transaction costs. Program budgets varied both within and across states. In Washington cases, program budgets were relatively high; Idaho and Montana exhibited moderate to low levels of program budgets with substantial intrastate variation in Montana. In Oregon, intrastate differences in program budgets were pronounced and varied from the high of Deschutes (ranked first) to the low of the Umatilla (ranked ninth) in the first implementation period.

The Deschutes remained the top program budget (with a slight decrease) across the two periods. The Salmon was an exception and moved from eleventh to fourth ranked in program budget, due to the 2008 Fish Accords agreement and the $7.6 million earmarked for water acquisitions in Idaho on the Lemhi and Pahsimeroi tributaries of the Salmon. The Upper Columbia and Yakima also increased their annual average program budget in period 2, while the weaker performers decreased, particularly in Montana where the Bitterroot program budget shrank by 60 per cent after hitting the wall in its water recovery. In short, performance has been uneven, both within and across states, and results from the 2008–10 period suggest that the gap between the stronger and weaker sub-basins appears to be growing.

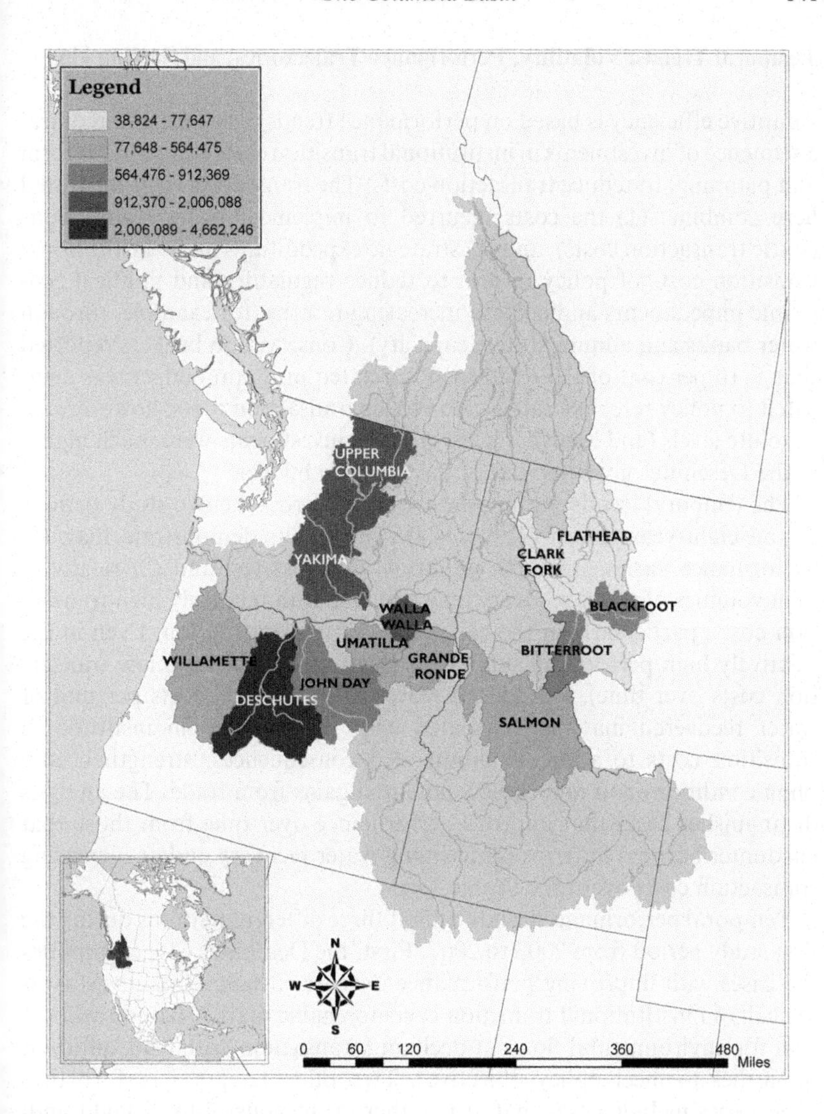

Note: This figure represents the program budgets (transaction-related expenditures) for two entities over the 2003-2007 implementation period: non-profit conservation brokers and state regulatory agencies in their central and field offices. The sub-basins were ranked and mapped using quantiles, as summarized in Garrick and Aylward 2012. Each quantile represents a 20th percentile interval, with darker shades indicating higher levels of each performance variable.

Figure 4.1c Spatial variability in program budgets (US dollars in $2007 dollars), 2003–2007

Temporal Trends: Volatility, Performance Trajectories, and Scaling Up

Adaptive efficiency is based on performance trends over time and requires a sequence of investments in institutional transition costs (for policy reform and planning) to reduce transaction costs. The transaction costs measured here combine: (1) the costs incurred to implement water transactions (static transaction costs); and (2) strategic expenditures in the institutional transition costs of policy reform to reduce regulatory and political economic impediments and achieve increasing returns (for example, through water banks and administrative capacity). Conservation brokers reported that 5–10 per cent of their transaction-related program budget was dedicated to policy reform strategies to reduce transaction costs; however, the absolute levels (and focus) of policy reform investments were much higher in the Deschutes given the size of the program budget.

The temporal trends capture the evolution across the two study periods for an eight-year record (Tables 4.4). The results demonstrate that: (1) performance varies as much within as across states; and (2) relatively high volumes of water recovery may coincide with relatively high transaction costs, particularly in the early years of implementation. Even in the relatively high performing sub-basins (high water recovery, low transaction costs over time), periodic increases in transaction costs per unit of water recovered may be associated with reinvestment in institutional transition costs to adapt to unintended consequences, strengthen enabling conditions and unlock new potential gains from trade. The analysis distinguishes cases that improve performance over time from those that encounter barriers in terms of declining water recovery and/or increasing transaction costs over time (Table 4.4).

Temporal performance trends exhibit three different patterns during the first study period from 2003 to 2007. First, the Deschutes case exemplifies the cases with improving performance over time; these cases invest strategically in institutional transition costs to enable market-based reallocation for environmental flows at declining transaction costs and sufficient quantities to meet policy objectives. Two other temporal performance trajectories include cases that were either (a) responsive to drought and/or (b) experienced enduring institutional barriers after initial implementation experience. The Washington cases are prominent examples of the drought-response trajectory whereby water recovery spiked and transaction costs declined during the 2005 drought year (declining from $20950/CFS in 2003 to $645/CFS in the 2005 drought year, before increasing to almost $7000/CFS in 2007), but residual institutional barriers tied to the lack of a water rights adjudication hindered implementation progress in non-drought years. Some of the Montana cases reflect the persistence of

Table 4.4 Performance trends in 10 most active sub-basins, 2003–2010

Sub-basin, State	Metric	Trend (2003–10)	Average Annual	Minimum	Maximum
Bitterroot, MT	Water Recovered		212	0	1 556
	Transaction Costs		$13 815	$76	$39 394
	Program Budget		$121 215	$47 369	$185 115
Blackfoot, MT	Water Recovered		98	0	431
	Transaction Costs		$17 133	$220	$61 661
	Program Budget		$117 760	$70 377	$173 457
Deschutes, OR	Water Recovered		330	104	580
	Transaction Costs		$3 870	$1 202	$7 630
	Program Budget		$890 183	$697 708	$1 160 010
Grande Ronde, OR	Water Recovered		29	0	96
	Transaction Costs		$5 524	$820	$17 438
	Program Budget		$103 674	$47 038	$122 199
John Day, OR	Water Recovered		57	0	232
	Transaction Costs		$8 241	$747	$27 128
	Program Budget		$167 148	$62 433	$251 899
Salmon, ID	Water Recovered		170	8	500
	Transaction Costs		$1 182	$374	$2 638
	Program Budget		$152 352	$90 617	$286 094
Umatilla, OR	Water Recovered		24	0	127
	Transaction Costs		$13 278	$681	$31 611
	Program Budget		$102 557	$45 937	$153 690
Upper Columbia, WA (Methow and Okanogan)	Water Recovered		64	0	229
	Transaction Costs		$30 905	$1 783	$159 904
	Program Budget		$335 823	$206 427	$533 300
Walla Walla, WA	Water Recovered		78	1	237
	Transaction Costs		$34 030	$1 392	$194 435
	Program Budget		$251 579	$132 334	$366 756
Yakima, WA	Water Recovered		133	16	684
	Transaction Costs		$8 393	$645	$20 950
	Program Budget		$428 060	$320 419	$585 334

Notes:
Note carefully that the average annual transaction costs per CFS is not the same as the average transaction costs for the full study period, for the reasons outlined above. A sub-basin with volatile recovery levels but stable budgets will see the annual average skewed higher than the average for the whole period. For example, in the Walla Walla and Upper Columbia, an individual year or two of very low water recovery levels combined with stable or growing program budgets, created outliers of exceptionally high annual values for transaction costs per cfs, while the remaining years were much closer to the longer term averages presented in Table 4.5.
Two years of continuous activity is required to establish a trend. All gaps in the trend analysis refer to periods without water recovery. Willamette trends were not extended for the 2008–10 period due to limited activity (within the QLE community). Clark Fork and Flathead sub-basins (of Montana) were excluded from both periods due to limited activity in the second period.
Annual transaction costs per cfs are not calculated for years without water recovery (it is technically impossible to divide by zero). These years are omitted from the calculation of the average of annual transaction costs per year, underestimating average annual transaction costs per cfs in sub-basins like the Bitterroot and Grande Ronde with multiple years without water recovery.

Table 4.5 Summary statistics for ten active sub-basins

(a) Total Water Recovered in Each Period (CFS present value)		
	2003–2007 (5yrs)	2008–2010 (3yrs)
Sub-basins, State	Total	Total
Bitterroot, MT	1,618	79
Blackfoot, MT	729	56
Deschutes, OR	1,278	1364
Grande Ronde, OR	101	100
John Day, OR	348	110
Salmon, ID	421	940
Umatilla, OR	181	12
Upper Columbia, WA (Methow and Okanogan)	106	408
Walla Walla, WA	550	74
Yakima, WA	902	165

(b) Transaction Costs per Unit of Water Recovered in Each Period (USD2007)		
	2003–2007 (5yrs)	2008–2010 (3yrs)
Sub-basins, State	TC/CFS	TC/CFS
Bitterroot, MT	$488	$2,341
Blackfoot, MT	$923	$4,779
Deschutes, OR	$3,649	$1,803
Grande Ronde, OR	$5,589	$2,647
John Day, OR	$2,623	$3,866
Salmon, ID	$940	$713
Umatilla, OR	$2,935	$24,004
Upper Columbia, WA (Methow and Okanogan)*	$13,383	$3,094
Walla Walla, WA	$2,492	$8,692
Yakima, WA	$2,225	$8,592

(c) Total Program Budget in Each Period (USD2007)		
Sub-basins, State	2003–2007 (5yrs)	2008–2010 (3yrs)
Bitterroot, MT	$789,699	$185,854
Blackfoot, MT	$673,228	$268,848
Deschutes, OR	$4,662,246	$2,459,219
Grande Ronde, OR	$564,475	$264,914
John Day, OR	$912,369	$424,817
Salmon, ID	$395,718	$670,746

Table 4.5 (continued)

(c) Total Program Budget in Each Period (USD2007)

Sub-basins, State	2003–2007 (5yrs)	2008–2010 (3yrs)
Sub-basins, State	Total	Total
Umatilla, OR	$532,403	$288,051
Upper Columbia, WA (Methow and Okanogan)	$1,423,977	$1,262,604
Walla Walla, WA	$1,369,912	$642,721
Yakima, WA	$2,006,088	$1,418,392

Excludes Willamette, Clark Fork and Flathead from period 1 due to limited activity in period 2 (as determined by the allocation of program budgets, and water recovery levels).
Zero annual values for water recovery in at least on year during period 1: Bitterroot, Grande Ronde, Upper Columbia.
Zero annual values for water recovery in at least one year during period 2: Bitterroot, Blackfoot, Grande Ronde, John Day, Umatilla.
Transactional activity includes state regulatory databases in period 1, but only the Columbia Basin Water Transactions Program database in period 2. In period 2, transactional activity was increasingly coordinated and cost-shared by the CBWTP in the selected sub-basins. To correct for any biases, the program budgets in period 2 were allocated to account for only the transaction-related expenditures for transactions financed by CBTWP. The transaction-related expenditures for all other transactions implemented by the relevant organizations were typically negligible and excluded.
All references to rankings of the sub-basins during the first period refer to the results presented in Garrick and Aylward 2012, which included all 13 sub-basins.

institutional barriers that limit potential for improved performance over time without reinvestment in market-enabling policy reform.

The time period for 'averaging' makes a big difference when interpreting performance trends, their volatility and implications. The spread is far wider for the average annual transaction costs than for the average transaction costs over the first five-year period. Average annual transaction costs across ten active sub-basins (Table 4.4) ranged from almost $1,200 to over $30 000, with the latter demonstrating the volatility from to year to year of sub-basins with patchy enabling conditions or intensive investment in institutional change. In the Upper Columbia, for example, transaction costs per CFS eclipsed $150 000 in 2006. Temporal performance trends in selected cases within Montana and Oregon are indicative of relatively low and high adaptive efficiency, respectively. The Bitterroot and Blackfoot achieved high water recovery levels and low transaction costs for the first study period. However, disaggregating the temporal trends into annual values for these two Montana sub-basins reveals that water recovery levels have declined and transaction costs are increasing. In some years, transaction costs have proven prohibitive and prevented water recovery; this trend is borne out by the 2008–10 period of analysis when both sub-basins had years without water recovery. Initial transactions used forbearance agreements that avoided administrative scrutiny and worked around the limitations of the prevailing institutional framework. Implementation efforts in western Montana shifted from a public agency (that is, similar to the Idaho government-run program) to private non-profits after legislation passed in 1995 enabled private leasing of instream water rights. As a result, staffing levels dedicated to environmental water transactions were lower in Montana than the other states during the study period. However, the state agency in Montana has proven important for sustaining high levels of performance, and initial transactional activity underscored the need for regulatory clarity and administrative oversight. The limited amount of state agency involvement partly explains the declining water recovery levels over time.

In Oregon, several of the cases mirror the Montana experience, with the majority of water recovery occurring in the first two or three years of the five-year term before 'hitting the wall' under prevailing institutional arrangements. The Deschutes Basin is a notable and illustrative exception. In the first study period, Deschutes achieved the second-highest water recovery levels (and the highest number of transactions by an order of magnitude) but also the third-highest transaction costs per unit of water recovered. The Deschutes had the highest program budget ($4.7 million) to cover transaction costs in the initial study period, invest in institutional reforms (water planning and banking) aimed at decreasing transaction costs over time, and generate sufficient financing to get to scale. Thus, the apparent contradiction of relatively high water recovery and high transaction costs is explained by the commitment of the Deschutes River Conservancy (DRC) to invest in the costs of implementing transactions while simultaneously bearing the costs of investing in: (1) communications, basin-level planning and policy forums to garner public support; (2) collaborative relationships with local irrigation districts; and (3) novel institutional water banking mechanisms and partnerships. For example, stakeholders developed and implemented a new regulatory program for exchanging existing senior surface water rights for new groundwater rights, representing a major policy reform that basin stakeholders worked through with the state during the study period. Tracking of trends over the five-year period shows a marked downward annual trend in transaction costs per unit of recovered water in the Deschutes. Compared with the other sub-basins, the annual transaction costs are relatively stable (with limited differences between the minimum and maximum values). These trends were consolidated in the 2008–10 period when the annual average transaction costs decreased. This reflects the pay-off in performance from continued investment in institutional reform and multilevel governance.

Interactions of Performance Variables and the Prospects for Adaptive Efficiency

Adaptive efficiency can only be understood through the interaction of variables and over time. Consider the three performance variables in terms of the combinations compatible with adaptive efficiency. Table 4.6 displays the eight logically possible combinations of outcomes, and comments on their feasibility and their implications for adaptive efficiency. The efficiency criterion identifies a standard for both a relatively high (high recovery, low transaction costs, sufficient budgets) and relatively low (the opposite) performance combination. If assessed over the longer term, two other combinations are relevant for both high (high recovery, high transaction costs, and sufficiently high capacity to cover the costs) and low values (the opposite, where recovery is low despite low transaction costs because of insufficient program budget capacity).

As Table 4.6 illustrates, transaction costs can be relatively high or low so long as institutional capacity is sufficiently high to cover and decrease the costs over time. Thus, high transaction costs are not automatically a negative for long-range effectiveness. The Deschutes and sub-basins of western Montana are illustrative. The Deschutes illustrates that relatively high transaction costs may be necessary during early periods of implementation when strategic investments in institutional transition costs (water banks, water planning, and so on) are needed to decrease transaction costs over time and reach water recovery thresholds for fish.

The typology in Table 4.6 also shows how the cases could be sorted in this scheme to gauge trajectories of performance over time. First, the high-performance combination of high water recovery, low transaction costs and sufficient program budgets appears to hold in the Willamette, Yakima and Salmon cases. Efficiency in these cases depends on prior investments in the transition costs of institutional change, and may require continued investment in institutional transitions in response to emerging barriers and opportunities. These prior investments in policy reform may cause transaction costs to be relatively high prior to the study period and/or borne by actors who are outside of the study scope but involved in water rights reforms that have spillover benefits for market-based environmental reallocation. Second, the longer-range concept of adaptive efficiency holds in the Deschutes, where high water recovery is combined with relatively

Table 4.6 Combinations of performance variables 2003–2007

Water recovery	Transaction costs	Program budgets	Cases
High	Low	Sufficient	Efficiency Salmon, ID; Willamette, OR; Yakima, WA
High	High	Sufficient	Potential for long-run adaptive efficiency Deschutes, OR
High	Low	Insufficient	Persistence of institutional barriers Bitterroot and Blackfoot, MT
High	High	Insufficient	Implausible
Low	High	Potentially sufficient	Persistence of institutional barriers followed by increasing recovery at declining costs per unit of water Upper Columbia, WA
Low	Low	Insufficient	Persistence of institutional barriers Clark Fork and Flathead, MT
Low	High	Insufficient	Inefficient Umatilla, OR; Grande Ronde, OR
Low	Low	Potentially sufficient	Path dependency Walla Walla, WA

Note: The Bitterroot, Blackfoot, Salmon, Willamette, and Walla Walla cases have intermediate values (that is, middle quantiles) for one performance variable. For example, the Salmon, Idaho, case has intermediate/low levels of program budgets (capacity) within the second-lowest quantile. The John Day was excluded, as it exists in the middle quantile for all three variables. Cases with 'potentially sufficient' levels reflect instances with moderate to relatively high program budgets and evidence of increasing returns (increased levels of water recovery with stable or decreasing transaction costs). The Walla Walla results should be treated with caution due to data quality gaps. Since 2010, the Umatilla has overcome institutional barriers, demonstrating that these categorizations are not static.

high transaction costs and also with the program budgets to cover and decrease those costs over time.

Other cases with high water recovery show less evidence of adaptive efficiency due to insufficient program budgets and therefore limited investment in the institutional transition costs of policy reform (rows 3 and 4 in Table 4.6). The Montana cases show high water recovery and low transaction costs but relatively low program budgets. As a consequence, institutional barriers persist and make it unlikely that the cases will improve performance over time unless prohibitive static transaction costs motivate efforts to invest in the institutional transition costs of policy reform. The final combination of performance trends with high water recovery, high transaction costs and low program budgets is logically impossible, and thus no such cases were found in this study sample.

The cases with low levels of water recovery are also informative. Without sufficient program budgets to address barriers to implementation and to foster market-enabling institutional change, the cases are likely to be trapped in a low-performance equilibrium. Therefore, even cases with initially low levels of water recovery during the five-year study period can be differentiated to gauge performance trajectories over time. The key difference is marked by those cases with high program budgets to cover the costs of getting to scale and to invest in institutional changes that will decrease transaction costs over time.

The Walla Walla and Upper Columbia of Washington reflect such cases. In the Upper Columbia, for example, the initial years of implementation were marked by low water recovery and high transaction costs. However, the sub-basin invested in relatively high program budgets, which contributed to increasing water recovery and decreasing transaction costs over time (as evidenced by the annual trend data, and the second period of analysis). The Walla Walla, however, demonstrates that high program budgets are not sufficient to improve performance over time unless investments in institutional transitions are effective. Thus, the design of institutional reform becomes a critical variable, and there may also be a lag in response of performance variables to institutional changes. The importance of design and potential for a lag may explain why the Walla Walla remains a relatively low performer in the second period of analysis from 2008 to 2010. It is during the latter period when the investments in multilevel institutional transitions were concentrated. Hence the payoff, in terms of increased water recovery and decreasing transaction costs, may still be pending even if the institutional design proves well matched to the local setting.

The Montana cases and the Umatilla and Grande Ronde of Northeastern Oregon, in contrast, fall into the low-performance trap due to inadequate investments in the transition costs of institutional reform; however, this trend has shifted since 2008, when Montana actors became involved in rule-making to address institutional barriers.

Explanatory Factors: An Analysis of Multilevel Water Rights Reform and Watershed Governance

The previous discussion illustrates the interdependence of multilevel water rights reforms and local, watershed governance. The former (multilevel water rights reform) is considered a necessary, yet insufficient, condition for long-term adaptive efficiency. The latter (local watershed governance) is needed to couple state-wide water rights reform with field-level institutions for monitoring, administration, planning and collaborative decision-making. The coupling of multilevel water rights reform and local, place-based initiatives is needed to manage water's multiple and interlinked non-market values and uses (including financing and stewardship of instream flows in overallocated basins).

Following from the discussion above, exclusive, enforced and transferable property rights to water are viewed as axiomatic (necessary) elements of market-oriented allocation policies (Anderson and Leal, 2001; Posner, 1986). Market-oriented water rights reform has three conceptual dimensions: exclusion, transferability and enforcement/administration (Table 4.7). This section operationalizes these concepts to classify the watershed cases. Each element of water rights reform and administration is determined by nested sets of rules at all three levels of action and analysis: operational, collective choice and constitutional. These elements of water rights reform involve polycentric combinations of local and state-level institutional change to achieve exclusion, transferability and adminstrative capacity as captured by the coding questions in the Appendix to this chapter. On this basis, it becomes possible to examine the sub-basins along scales (from 0 to 1) of institutional design based on an equally weighted average of several attributes, and to track change by comparing these values across multiple snapshots in time (corresponding with each study period). Here I focus principally on the first study period.

Performance Trends and Institutional Change

How do performance trends relate to institutional change at multiple scales? The relationship between water rights adjudication and performance offers insight into this question. Oregon and Idaho are fully adjudicated in the sub-basins within the study area through processes that entailed costly litigation and field-level monitoring and enforcement capacity. Adjudications in Montana and Washington remain incomplete, although costly investments in adjudications have occurred in pockets within these states, such as the Yakima in Washington. The completion and enforcement of a surface water rights adjudication is expected to contribute to relatively high levels of performance when demand for environmental water recovery is high, as evidenced by the Deschutes and Salmon examples. The lack of adjudication is expected to constrain performance due to incomplete property rights, which helps explain the experience in the Upper Columbia of Washington.

Two sets of outliers are instructive: (1) those without adjudication but

Table 4.7 Institutional design attributes: concepts and measures

Water Rights Reform Category	Attributes
Exclusion	• *Basin closure* • *Surface water adjudication* • *Instream flow rule* • Groundwater mitigation rule • Transboundary status
Transferability	• Contract diversity • Contract duration • Adoption of instream transfer/leasing rule • Revision of instream transfer/leasing rule • Formal administrative processing timelines • *Water bank* • History of contested case hearing
Enforcement/ administration	• Watermaster (agency official) for field-level administration • Measurement and enforcement schedule • Conflict avoidance and resolution mechanisms
Equally weighted score for listed elements	*Partial score is possible for italicized sub-attribute* • policy reform or planning element can be initiated (0.25) or complete (0.5) • geographic extent can be partial (0.25) or complete (0.5)

with high levels of performance across the three measures; and (2) those with adjudication but with low levels of performance. The cases in the former category (for example, Bitterroot) and latter category (for example, Grande Ronde) illustrate two essential points about multilevel collective action needed for effective institutional reform. The Montana cases (in the first category) recovered high levels of water despite the absence of water rights reform. After the initial deals, the Bitterroot and Blackfoot sub-basins exhausted the potential for deals with low transaction costs within the prevailing institutional setting and found it difficult to build on early success to get to scale. These implementation challenges prompted efforts in 2008 to develop new rules and administrative pathways for environmental leasing to reduce barriers to trade. Thus, an adjudication and/or reinvestment in the transition costs of policy reform may prove necessary to achieve adaptive efficiency. Although changes in 2008 proved insufficient to unlock immediate improvements in institutional performance during the 2008–10 period, the western Montana region simultaneously restructured its implementing body when the Clark Fork Coalition, a

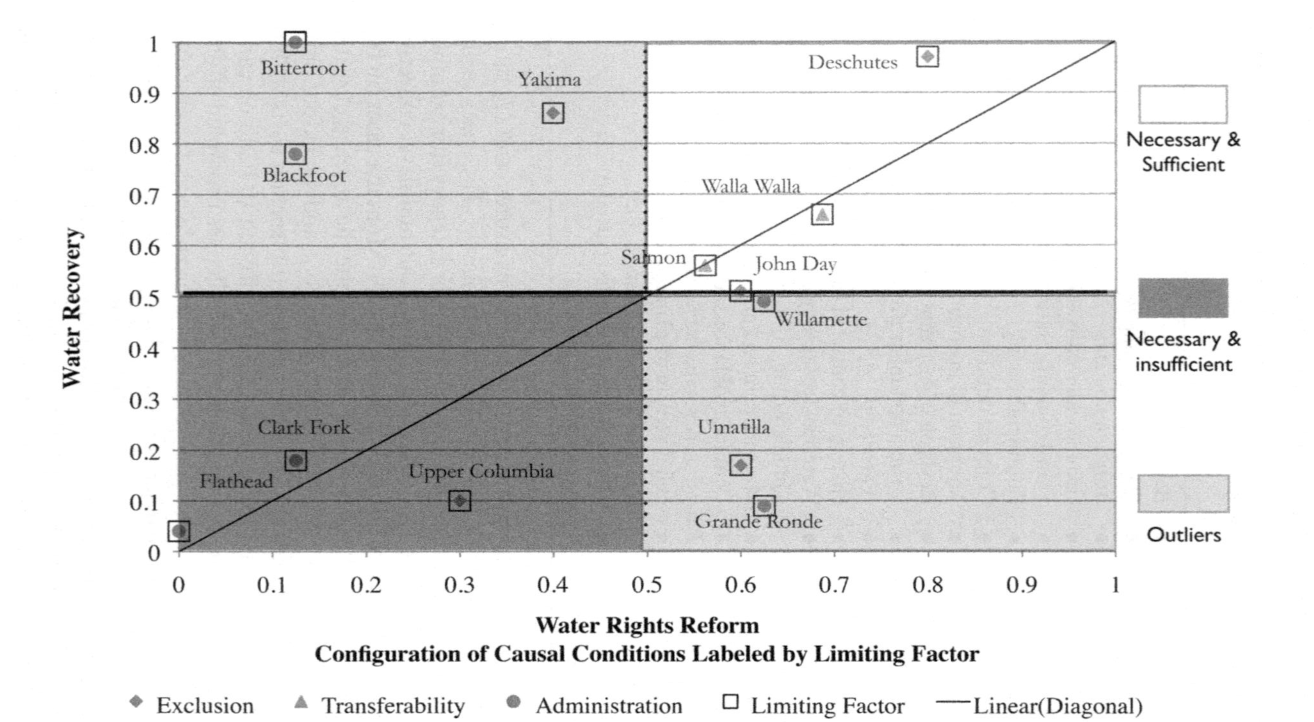

Note: Following the coding scheme listed in Table 4.7 and summarized in the Appendix. For full details of the coding scheme, see Garrick (2010).

Figure 4.2 Water rights reform and water recovery

regional body, acquired the Montana Water Trust, a statewide organization. The localized capacity and horizontal linkages with water users, field enforcement officers and other stakeholders may prove important to work around deficiencies in the institutional framework, and build the trust and collective action to address these institutional gaps in the future.

The Grande Ronde and Umatilla (in the second category) lack water recovery despite high levels of water rights reform (that is, via the presence of adjudication). These cases demonstrate that adjudication may prove necessary but insufficient for long-run efficiency unless such water rights reform is coupled with governance reforms by sub-basin actors to cooperate in strengthening enabling conditions (for example, water banking and reverse auctions and addressing unintended consequences of water transactions on a range of stakeholders). The Grande Ronde, for example, involved a permanent transaction that proceeded without consultation with the broader basin-wide community (Neuman, 2004; Pilz, 2006). The transaction triggered a protest from upstream and downstream water users that led to a costly, six-year conflict resolution process in administrative courts; irrigators rallied against the transaction and organized political resistance that threatened to unravel Oregon's legal and regulatory framework for water transactions (Neuman, 2004). The Grande Ronde case demonstrates the benefit of linking water rights reforms with basin governance efforts that promote local decision making over water allocation and conservation planning buttressed by complementary higher-level mandates (for example, state water planning, federal Endangered Species Act) to reinforce local initiatives. By contrast, in the Deschutes and Upper Salmon cases, water rights and multilevel basin governance conditions combined to improve performance over time. In both instances, water rights reform proves necessary but insufficient for adaptive efficiency.

Sustained performance depends also on watershed governance institutions to address vertical and horizontal coordination challenges due to the pervasive externalities and interdependencies associated with market-oriented environmental water allocation. The principles of subsidiarity (retaining authority and capacity at the lowest level possible) and complementarity (ensuring higher-level institutions for externalities and economics of scale that span political boundaries) are achieved through institutional mechanisms sometimes described as cross-scale linkages – the interdependency of two or more collective bodies to carry out governance functions (Heikkila et al., 2011).

Using the systematic coding of water rights reform characterized above, it is possible to categorize the sub-basins into quadrants based on multilevel water rights reform and water recovery, both on a normalized scale from 0 to 1 (see Figure 4.2). This categorizes the cases into those with evidence of adaptive efficiency for which high levels of water rights reform are necessary and sufficient for high levels of water recovery (top right quadrant); those with evidence of path dependency (bottom left quadrant); and those with the outlier conditions described above: (1) low hanging fruit: high water recovery, low levels water rights reform; and (2) collaborative deficits: low water recovery despite high levels of water rights reform. These quadrants are unpacked in the brief analysis of the different types of cases:

- Top right quadrant: the Deschutes miracle.
- Top left quadrant: the Montana exception, 'Low-Hanging Fruit'.
- Uneven performance and the promise and peril of the surgical strike, the challenge of 'collaborative deficits'.
- Balancing central and local capacity: the Washington experience.

The Deschutes miracle: coupling multilevel water rights reform with local collective choice bodies[13]

The Deschutes River of central Oregon spans more than 10000 square miles, parts of five counties, and the second-largest area of any river basin in Oregon (Aylward and Newton, 2006). The Deschutes River became fully allocated in the early 1900s (Aylward and Newton, 2006; Hubert et al., 2009). In the 1990s and early 2000s, new growth trends lent urgency to reallocation pressure from cities and unmet environmental water needs. Allocation stress caused by new growth led to the development of state and local rules, local collective choice venues for banking and water rights administration, and operational capacity for transactions among institutional sellers (irrigation districts), buyers (Deschutes River Conservancy), and field-level regulatory officials (watermasters). Growth in residential development led to the conversion of agricultural acreage where land use shifts targeted interest in acquiring water rights with high priority. The Pelton Dam impeded anadromous fisheries and thereby limited the influence of the Endangered Species Act on demand for environmental water. However, the 1987 Instream Water Rights Act had converted many established minimum perennial streamflow reservations into instream water rights, and new growth pressure spurred planning efforts that integrated instream flow needs among the major demand drivers alongside water for cities and agriculture (Aylward and Newton, 2006).

Federal and local processes complemented and reinforced the levels of exclusion, transferability and administration achieved through ongoing statewide water reforms affecting the Deschutes, often in response to growing population. The local and federal overlay demonstrates the role of cross-scale linkages in water rights reform within polycentric institutional settings. The Deschutes River experience demonstrated that statewide reforms are necessary but insufficient to reach high recovery levels and implementation budgets at low unit costs. The Deschutes experience suggests that it is necessary to invest in local collective choice and operational rules and to link with federal authority and resources. The development of local collective choice institutions proves costly in the stage of policy reform and implementation captured in the 2003 to 2007 snapshot.

The Deschutes Basin Working Group traces its genesis to a collaboration formed by tribal, environmental, irrigation and federal interests in 1992 to resolve long-standing tribal claims (Moore et al., 1995). By 1996, the Deschutes Working Group had framed the problem of allocation stress across human and environmental needs and had evaluated the potential institutional responses, including 15 ongoing planning processes and studies and 18 existing institutions (Moore et al., 1995: 112). This analysis reviewed four alternatives for coordinating these efforts to reform the institutional framework to enable water transactions as a strategy for shifting water in response to competing and evolving needs. The alternatives were: (1) private, non-profit; (2) quasi-governmental institution; (3) corporation under federal law; and/or (4) federal–state–tribal compact (Moore et al., 1995: ES-5). The resulting institutional change led to a private non-profit with quasi-governmental attributes, including federal authorization under the 1996 Oregon Natural Resources Conservation Act. This institutional choice stems in part from the comparatively low transaction costs of collective choice for a less formal entity developed at the local level instead of costlier compact negotiations at the federal level.

The establishment of the 1996 Deschutes Basin Working Group triggered a decade of collective choice and operational-level institutional change. In 1998, the Working Group negotiated its first lease and designed and engineered the first conservation project to enhance instream flows through efficiency savings along the canals and laterals serving the irrigation districts in the region. The federal influence of the US Bureau of Reclamation reservoir system and the local role of irrigation districts as large group water rights owners encouraged efforts to build on individual transactions by institutionalizing rules, processes and capacities for increasing water transactions to reach scale. By 2002, hydrologic studies had led to the development of a groundwater mitigation rule that built demand for instream transactions to offset needs for new growth, and the Deschutes Water Exchange formed a local collective choice venue to facilitate the creation of and access to mitigation credits by acquiring and retiring existing water rights to satisfy emerging demands.

In parallel, the major buyers, sellers, brokers and regulatory entities institutionalized their collective choice and operational capacities in the

Deschutes Water Alliance comprising the Confederated Tribes of Warm Springs, the Deschutes Basin Business Council with eight irrigation districts, the Central Oregon Cities Organization and the Deschutes River Conservancy. The Alliance convened a 2006 summit to plan for long-range water demands. Under this umbrella of overlapping, multilevel collective choice processes and rule-making in support of market-based allocation strategies, the Deschutes River Conservancy rapidly bolstered both water recovery levels and implementation budgets. The set up costs of building multistakeholder support boosted transaction costs, essentially internalizing the costs of 'being local' (building local buy-in) into the program's design and implementation rather then depending on independent or loosely coupled initiatives to secure stakeholder support. Its board of directors and bylaws establish positions to ensure representation of all of the key interest groups and local, state and federal agencies.

In short, water rights reform is costly but necessary to achieve steady gains in market delivery of environmental flows over the long term. The trends analysis above (see Tables 4.4 and 4.5) demonstrate a marked decline in transaction costs over time and far less volatility than many of the other sub-basins as the benefits of upfront collective choice institution-building contributed to the scale and adaptive efficiency of meeting long-range water needs for environmental flows. As of 2012, the Deschutes River case had achieved a portfolio of temporary and permanent acquisitions to restore 223 cubic feet per second and protect 330 miles of streams (DRC 2012). The DRC's latest annual report (2012) highlights an increasing proportion of water protected permanently, and key staff have remained with the organization for a decade.

The Montana exception: the limits to free market environmentalism without multilevel water rights reform[14]

The Montana cases vividly illustrate the logic and limits of market-oriented water rights reforms. The Montana framework for water rights reform has proved patchy and incomplete due to partial exclusion, transferability and administration levels, as noted above. Western Montana contains four sub-basins in the Columbia River – the Bitterroot, Blackfoot, Flathead and Upper Clark Fork – none of which contains anadromous fish due to passage restrictions at the Thompson Dam. Thus, the drivers for instream leasing center on threatened resident Bull Trout, and since the early 2000s the development of groundwater supplies for new growth has aligned irrigator and fishery interests in preserving surface flows for farms and fish, respectively.

The Montana cases exhibit relatively low degrees of market-oriented water rights reform along all three dimensions of exclusion, transferability and administration. While the Clark Fork and Flathead reach low recovery and implementation budgets, performance in the Bitterroot and Blackfoot during the initial implementation period (2003–07) implies that it is possible to recover large volumes of water for the environment without market-oriented water rights reform. As noted above, the Painted Rocks deal in the Bitterroot was an important achievement, but failed to build toward additional recovery. As such, the summary measures of performance are potentially misleading and exaggerated.

Unlike the Deschutes River, where water recovery and average transaction costs fluctuate within a relatively narrow range ($1202–$7630 for annual transaction costs and 104 to 580 per year in present value CFS acquired in the first study period), the Montana cases demonstrate how the five-year cumulative average costs and total water recovery levels are misleading. The long-term averages obscure the trends within that period and mask the importance of multilevel collective action to undertake the market-oriented water rights reform needed for sustained success. The Blackfoot and Bitterroot demonstrate large, proof-of-concept progress in 2003 and 2004 before reaching the limits of 'low-hanging fruit'. In 2003, the Blackfoot and Bitterroot notched 430 and 1556 present value CFS at a negligible transaction cost of $220 and $76 per CFS, respectively. Trout Unlimited secured path-breaking transactions in both the Bitterroot (Painted Rocks) and Blackfoot (a series of irrigation efficiency projects building on early innovation by Montana Department of Fish, Wildlife and Parks). However, by 2006, annual recovery levels had plummeted to almost negligible background levels, although implementation budgets remained constant or increased, leading to peak average annual transaction costs of $61 660 in the Blackfoot (2006) and $25 116 per CFS in the Bitterroot (2005) – a flatlining trend that continued into the second implementation period and included several years without any deals. The Bitterroot did not add any CFS in 2007. Its average transaction costs were therefore null for implemented deals even as its program budgets remained positive and amounted to enough to support a staff member working in the sub-basin. This demonstrates Cole's (2002) point that actual transaction costs diverge from potential transaction costs in important ways, particularly in cases where the actual transaction costs of zero hint at the prohibitively high transaction costs.

The Montana cases therefore present the classic evidence of an exception that proves the rule. Initial success proved fleeting in the absence of market-oriented water rights reform, which demonstrates the necessity of multilevel collective action to achieve exclusion, transferability and enforcement for long-term adaptive efficiency. The programs in the state only included non-profit budgets (unlike Oregon, Washington, and Idaho where state regulatory bodies dedicated significant budgets to regulatory and monitoring functions in the transaction cycle). Therefore, the limited administrative contributions to transaction costs lowered the average costs for initial deals. Reduced transaction costs came with a price: insufficient administrative capacity and transfer rules to enable long-range growth in transactional activity. Exclusion remains patchy. However, basin closure and localized verification and validation of water rights in places such as the 'Tin Cup' of the Bitterroot show that exclusion is rarely the limiting factor in market-oriented water rights reform. Instead, rules constraining transferability and limited administrative capacity affect the trajectories of reform in Montana's sub-basins. Western Montana's sub-basins suffered from two types of institutional deficits which typically coincided: (1) inadequate local layers of water rights reform (sub-basin rules, administration) to buttress the formal water rights institutions at the state level (adjudication, transfer rules); or (2) missing linkages between local watershed institutions and wider water rights reforms. While the Deschutes fused local collective choice bodies (Deschutes River Conservancy, among others) with state and local water rights reforms, initial transactional efforts in Montana proceeded only in areas where problems of exclusion, transferability and administration could be solved in an ad hoc (localized, temporary) fashion.

The deficits in formal administrative capacity and transferability channelled initial efforts to geographic contexts where only a single landowner or small group could enter into a forbearance agreement. The vast majority of transactions in the Montana cases were examples of such water use agreements in which the range of users affected by and bound by the proposed transaction are limited to the contracting parties, in which the conservation benefit is unaffected by upstream or downstream users. This context limited attention to habitat connectivity projects involving senior water rights holders near the confluence of a tributary and the mainstem. Once these low-hanging fruit were plucked, the Montana cases have since attempted to expand into formal administrative changes that would convert forbearance agreements into recognized instream leases with power to encumber other users and uses with junior priority. This transition has encountered the obstacles of patchy and incomplete water rights reform, as well as the inability to link effectively with concurrent collective choice institutions for watershed planning and restoration. The Blackfoot provides an illustrative example in this regard because parallel governance efforts in collaborative decision-making have protected up to 60 CFS per year through the Blackfoot Challenge Drought Task Force, a voluntary but non-market (that is, non-compensated) agreement

(Blackfoot Challenge, 2008). In sum, a voluntary uncompensated reallocation of 60 CFS occurred during drought, while, to paraphrase Neuman (2004), bank accounts for market-based environmental water acquisitions remained full, as it was too costly to spend the money.

In sum, the performance of the Montana cases underscores the paradoxical necessity of market-oriented water rights reform to maintain and improve performance levels over time and, ultimately, to scale up transactional activity and implementation resources to desired levels at the sub-basin scale. Ongoing reforms and collaborative processes, particularly in the Blackfoot, Bitterroot and wider Upper Clark Fork region, have worked to build on early transactions to enhance the transferability and administrative capacity for larger-scale outcomes. In 2009, the Montana Department of Natural Resources and Conservation (DNRC) continued efforts to develop new rules to govern administrative change procedures, while implementing entities forged tighter links between concurrent collective choice processes in the Bitterroot and Blackfoot sub-basin planning and related activities. Evidence of the renewed emphasis on local collective choice and operational rules-in-use appeared in the efforts by both non-profit brokers – Trout Unlimited and Montana Water Trust – to influence the administrative change guidelines for instream flows. The Clark Fork Coalition acquired Montana Water Trust and subsumed its functions within a watershed-oriented scope of integrated restoration efforts, illustrating the role of horizontal linkages between transactional approaches and wider collaborative governance efforts. Despite the initial challenges in Montana, the state regulatory body has become increasingly engaged since the 2009 period, reestablishing its role in water transactions since the transition from public to private leasing programs in the 1990s. Although transactional activity remains low in the Montana sub-basins today, important and creative deals continue to get done, as illustrated by the almost $600,000, 12 cfs permanent deal brokered by Trout Unlimited-Montana in the Nevada Spring Creek in 2013 to transfer the right for instream use to be held by the Montana Department of Fish, Wildlife and Parks.

Uneven Performance: State-level Enabling Conditions, Differing Local Capacity, and the Promise and Peril of the Surgical Strike[15]

The most complex cluster of cases present moderate levels of performance across all criteria – water recovery, transaction costs and implementation budgets – for the first study period as well as relatively high levels water rights reforms. Although each achieves roughly similar levels of market-oriented water rights reforms, the constellation of outcomes differs markedly across all three cases when assessing the trajectory over time. In Oregon, the John Day and Grande Ronde exist side by side geographically and share many similarities. Like the Upper Salmon, these eastern Oregon cases represent sub-basins where agriculture dominates natural flow water systems (rather than regulated upstream reservoir storage), and the environment represents the primary competition with agriculture for water in fully allocated reaches. Urban pressure has not driven environmental water acquisition efforts in these areas, but endangered species values are exceptionally high, with anadromous species and habitat concentrated in these tributaries. At the governance level, each watershed has progressed evenly in terms of formal and informal water rights reforms. Each has also developed local collective choice bodies for watershed planning and received designations as 'model watersheds' under the Northwest Power and Conservation Council fish and wildlife program, demonstrating the strength of local institutions.

The performance outcomes vary despite otherwise similar water rights conditions in ways that illustrate the hidden costs and benefits of a 'surgical strike'[16] approach to environmental water acquisition in remote rural settings like these. The surgical strike describes a project-based approach to environmental water transactions where deals are isolated and a single, well-targeted project can make a substantial difference; without requiring explicit coordination with wider restoration and collective choice processes. This approach was common in the early phases of implementation to demonstrate proof of concept and verify that water acquired through market transactions led to a net increase in streamflows and contributed to habitat outcomes. The approach finds traction in statewide organizations whose scarce resources lead to the establishment of priority areas in rural parts of the state where a single, small transaction could generate a disproportionately high ecological value or potential to seed further transactions. Such projects often involve a five-hour drive from a capital city (Portland and Seattle) for due diligence, negotiation and oversight, for example in the distance between the Portland headquarters of the Oregon Water Trust (now The Freshwater Trust) and the John Day or Grande Ronde. The approach does not imply a competitive or adversarial process relative to the other case studies above, but rather a more targeted and limited involvement in local collective choice processes.

Within the surgical strike approach, institutional diversity reigns, as much within states as across them. The John Day exhibits moderate levels of water recovery, transaction costs and program budgets. The Grande Ronde, on the other hand, only achieves low transaction costs by virtue of its focus on the low-hanging fruit in projects proceeding outside of the formal administrative system. The Upper Salmon is a pocket of market-oriented water rights reform in Idaho. Its combination of state-level legislative reform and local capacity has contributed to good performance: high water recovery, low and declining transaction costs, and increasing program budgets.

The three cases demonstrate different ways of pursuing local transactions from afar, and performance levels reflect the strengths and weaknesses of the different methods used to bridge from statewide legal and administrative frameworks to the local context, where local collective choice by key stakeholders, project design and implementation, and operational decisions about water use and distribution take place. In Oregon, with the John Day and Grande Ronde, the Oregon Water Trust experienced both the promise and pitfalls of the surgical strike approach. During the initial implementation period from 2003 to 2007, the John Day of eastern Oregon achieved relatively high water recovery levels and implementation budgets at low transaction costs.

The project officer from the Freshwater Trust (then the Oregon Water Trust) established strong collaborative ties with local collective choice bodies and operational capacity developed by the basin's Soil and Water Conservation District and field-level watermaster of the state water agency.[17] These performance outcomes proved fragile due to the dependence on a single project officer to couple water rights reform and local collective choice bodies. In other words, the departure of a single staff member threatens to unravel the collaborative relationship between the conservation broker and key local and informal water rights institutions and watershed governance bodies. The Grande Ronde has also proven a font of innovation and path-breaking transactions, such as the minimum flow agreement on the Lostine River that netted 15 CFS and proceeded as a forbearance agreement outside of the formal system for water transfers. However, despite the ecological value and advanced development of the Grande Ronde model watershed program, the inability to link statewide reform and field-level water rights reform with local institutional development has constrained long-range progress in the Grande Ronde until relatively recently, when a sustained and targeted focus on the Catherine Creek region has supported an expanding portfolio of water acquisitions.

The Lemhi River of the Upper Salmon exists apart from the Oregon cases and reflects the importance of multiple layers of market-oriented water rights reform and institutional capacity at the local and state levels. In response to the threat of endangered species enforcement for the Chinook Salmon, the state legislature authorized a minimum stream flow of 35 CFS in the region and formed a unique natural flow rental pool to facilitate transactions to meet this goal. In 2003 and 2004, the Idaho Department of Water Resources forged linkages with the Upper Salmon technical workgroup emerging out of a mid-1990s NPCC model watershed project, and also enlisted the local watermaster for extension. The involvement of the state agency as the primary broker reinforced the linkage between state-wide and field-based layers of market-oriented reform and ensured that local processes would benefit from efficient administrative procedures at the state level. As a result, water recovery levels have steadily increased and transaction costs have stabilized at relatively low levels, as depicted in the trend analysis above. In 2008, the progress in the Lemhi led to a ten-year memorandum of agreement under the Bonneville Power Administration fish and wildlife recovery efforts, with $7.5 million in funds for implementation and acquisition. The strong linkages and layering of market-oriented water rights reform has ensured continued progress along all dimensions of performance and, despite the initial limitation on permanent transactions and irrigation efficiency projects under Idaho's formal water code, the program has proceeded to extend and expand its transactional activity for longer periods and over wider geographic scope across the Upper Salmon. The Deschutes and Upper Salmon demonstrate that state-wide market-oriented water rights reform is necessary but insufficient and must be coupled with local collective choice bodies and operational capacity as either a follow up (Deschutes) or a prelude (Salmon) to state-wide institutional reforms. The Umatilla of Oregon (since 2011) and Upper Columbia of Washington (since 2008) have embarked on institutional reforms to address collaborative deficits and scale up water recovery levels in rural sub-basins, with similar geographic conditions as those discussed in this section.

The Washington experience: institutional diversity to balance state and local, public and private capacity

Performance outcomes vary within Washington and over time due to piecemeal progress in water rights reform and uneven institutional capacity for collaborative planning, water banking, joint monitoring and field-level enforcement. The Yakima, Walla Walla and Upper Columbia Cascade regions provide the ideal contrast for exploring how market-oriented water rights reform and watershed governance interact in the design and performance of market-based environmental water acquisition. The experiences of these sub-basins in phase 1 (2003–07) diverge. The fortunes of the Walla Walla (from high performance to low) and Upper Columbia (from low to high) shifted between the two periods. This demonstrates that efforts to enhance institutional diversity by coupling market-oriented water rights reform and multi-layered watershed governance capacity can reduce transaction costs, increase water recovery and crowd in investment needed to get to scale.

The Walla Walla is noteworthy because of its unique characteristics. As the only watershed case spanning two states in the study area, interstate cross-scale linkages are needed to coordinate institutional reform and implementation in Oregon (upstream) and Washington. Spanning 1758 square miles, almost three-quarters of the basin falls in Washington (Siemann and Martin, 2007). Moreover, experiments in watershed governance and local flexibility have focused on the Washington side, in part because legal and administrative frameworks still pose formidable barriers to protecting water rights retired for environmental flows in upstream Oregon (Siemann and Martin, 2007).

In the first phase, markets for environmental flows in the Walla Walla have relatively high water recovery levels, low transaction costs and high implementation budgets. The Walla Walla has not attained the status or scale attained in the Deschutes. The region has undergone market-oriented water rights reform with both state- and local-level legal and rule changes; this has been closely related with the development of collective choice venues and cross-scale linkages for watershed planning. These linkages preserve local capacity while capturing state and federal resources for local reform culminating with the 2008 instream flow rule amendments and establishment of a groundwater mitigation bank. Like the Deschutes, the Walla Walla's progress is founded in part through its unique experiment in watershed governance under the Walla Walla Water Management Initiative begun in the mid-2000s. Unlike the Deschutes, where tribal negotiations and hydropower relicensing provided the external triggers for institutional reform, the Walla Walla was galvanized by the Endangered Species Act (ESA). The ESA cast a shadow of coercion that provided the impetus for its experiments at the confluence of watershed collaborative and market-oriented models.

The listing of the Bull Trout and Middle Columbia Summer Steelhead in 1998 and 1999 was followed by a threat of enforcement action in 2000. Under these conditions, the three major irrigation districts developed a quasi-mandatory reallocation on the Oregon and Washington sides of the border through a negotiated settlement of up to 27 CFS in Oregon and 19 CFS in Washington (Siemann and Martin, 2007).

Against this backdrop, the past decade witnessed an alignment of collaborative and market-based water allocation reforms. By the time the 2008 instream flow rule and mitigation bank formed, the Walla Walla was poised to scale up. That potential has not been realized, however, in phase 2. This progress has yet to deliver major increases in water recovery levels, but it has established fledgling collective choice institutions for water banking, watershed planning and a state-wide water trust with a strong field presence.

The Walla Walla experience exhibits the Janus face of subsidiarity and complementarity. Local capacity is necessary but insufficient; a multilevel approach is needed. By the second study period (2008–10), progress in the Walla Walla lagged due to limited financing, changing land use trends (in part due to the housing crisis limiting 'demand' for groundwater mitigation) and, by extension, comparably limited competition between consumptive and non-consumptive uses to facilitate larger-scale outcomes achieved in the Deschutes. Both of the other cases in Washington – the Yakima and Upper Columbia – underscore the importance of cross-level interactions between subsidiarity and complementarity in polycentric watershed governance; state and federal efforts have developed (initially) in the absence of local capacity, at least in the first phase (Singleton, 1998; Weber et al., 2005). Washington is noted for both its abiding commitment to salmon recovery and market-oriented reform at the state level, and its equally stubborn resistance among local rural communities in targeted watersheds due to rigid application of the prior appropriation doctrine and perceived regulatory risks during the administrative process for environmental water transfers. Notably, the Walla Walla experiment represented a deliberate attempt to 'flow from flexibility' in deviating from state-wide regulatory frameworks to assuage the fears of irrigators concerned that participation in market-based environmental flow projects will lead to the relinquishment of their water rights. (Siemann and Martin, 2007).

The two other cases in Washington bookend the experience with linking federal and state initiatives with local capacity and collaborative efforts to develop and implement market-based environmental water allocation. In the first phase, the Yakima represented a successful case in terms of performance, while the Upper Columbia initially lagged in performance due to lingering resistance to ESA enforcement. However, its transition in the second phase was striking; it was one of the sub-basins where transaction costs decreased from period 1 to 2 and its other performance variables exhibited signs of adaptive efficiency in their trajectories after being constrained by path dependency and vested interests in the first period.

In the Yakima, a federal water storage project, well-organized tribal government and an infusion of federal dollars together created strong state and federally led initiatives and enabled progress despite uneven water rights reform and a water rights adjudication underway since the 1970s. The Washington state government's Trust Water Program was piloted in Yakima via state legislation due the special federal character of the basin and its storage infrastructure. These powerful external mandates and financing sources proved sufficient despite limited local capacity. The Yakima lacks a general-purpose, multistakeholder watershed organization, although a watershed planning unit has been formed under the auspices of the 1998 Watershed Planning Act. A primary vehicle for establishing local capacity and encouraging subsidiarity has been a working group for the Yakima water bank. The heavy reliance on state and federal authority and capacity has meant that the Bureau of Reclamation exerts substantial influence over the design and management of the project, which makes this case more similar to the Upper Snake program in Idaho, which was excluded from the study due to the minimal role of local institutions. Tacit recognition of the need for local capacity and cross-scale linkages has come in the form of the Washington Water Trust's field office in the region. Drought crises have temporarily reduced legal and administrative barriers in water rights institutions and revealed the importance of local capacity to pursue allocation flexibility.

While the heavy reliance on state and federal authority and financing has not hampered the Yakima (which has protected the second-highest level of water instream next to the Deschutes over the five-year period in phase 1), the Upper Columbia lacks the federal overlay associated with the Bureau of Reclamation's Yakima project. As a consequence, the state government is a prime influence and source of financing, along with the Bonneville Power Administration, which is active in all of the case studies.

The sub-basin struggled to gain traction until recent progress since 2008. ESA enforcement action in 2002 polarized the community. Even before this influence, the state attempted its first Trust Water project in the Methow of the Upper Columbia and encountered challenges in gaining approval for the project due to opposition from other water users. Therefore, despite the strong federal and state-driven acquisition efforts connected to the region's high ecological priority and financing, the Upper Columbia remained slow to develop. In 2006, the Washington Rivers Conservancy (now reformed as Trout Unlimited – Washington Water Project) established an office in the region and began to coalesce stakeholders from different parts of the Upper Columbia to build the horizontal linkages needed to bolster local capacity. Since the group's formation, the Colville tribes have also become active under the 2008 Fish Accords developed by the Bonneville Power Administration to partner with the tribes in designing, financing and implementing programs. The shifting fortunes experienced in the Upper Columbia demonstrate the potential for sequential investments in institutional transitions to achieve the combination of subsidiarity and complementarity needed to reduce transaction costs and achieve adaptive efficiency.

CONCLUSIONS

This chapter has examined transaction costs and institutional performance in market-based environmental water allocation. It developed and tested a transaction costs framework to evaluate market-based environmental water allocation. Performance – water recovery, transaction costs and institutional capacity (program budgets) – varied as much within states as across states.

Five chief lessons emerged. First, institutional responses to public goods failures in water allocation are costly; the transaction costs involve important interactions between transition and transaction costs, which are therefore difficult to separate in accounting frameworks. Second, context matters but is not determinative; differences in context favor nested governance arrangements comprising field, state and federal institutions matched to local conditions which can vary as much within states as across them. Third, clear methodological categories and typologies are needed to define and assess transaction costs and institutional performance through all stages of policy design and implementation. The methodological issues include defining which set of actors are incorporated, particularly to account for and explain the public transaction costs required to deliver water-related public goods at socially desired levels through market-based mechanisms. Fourth, performance varies spatially within states as well as across them. The performance trends suggest that high levels of water recovery depend upon adequate financing for investments in institutional transitions to couple water rights reform and watershed governance capacity. Such investments explain why performance varies within states in response to ongoing state- and watershed-level policy reforms, such as water banking and water allocation planning in the Deschutes and Salmon examples. These conditions enhance the likelihood of achieving adaptive efficiency, which was defined as sufficient levels of water recovery at low and/or declining transaction costs, with adequate institutional capacity (program budgets) to get to scale. Finally, temporal performance trends within cases also demonstrate the importance of reinvesting in the institutional transition costs of policy reform, especially during early periods of implementation when initial water recovery efforts either require relatively high transaction costs or are only able to proceed at low transactions costs because implementation efforts target transactions that are not affected by persistent institutional barriers, such as the lack of an adjudication or administrative rules. Adaptive efficiency requires a longitudinal perspective and time series data to determine whether performance is likely to achieve scale. Future research should incorporate public transaction costs and institutional performance into a comprehensive and longitudinal assessment of benefits and cost. Such assessment will evaluate institutional design and track the factors contributing to effective market-based water policy.

NOTES

1. A range of complementary funding from state programs, non-profit fund raising and mitigation projects by cities and energy utilities have piggybacked on federal expenditures.
2. See Coase (1960) and Demsetz (1967) for earlier and narrower definitions of transaction costs.
3. Defined as net increases in flow rates through water rights acquisitions to retire historic water uses and rededicate the historically consumed water for ecological purposes instream.
4. It should be noted that implementation cost per unit of water is derived as an average cost (total transaction-related program budgets divided by the net present value of water recovered in CFS). Average transaction costs metrics are prevalent in water-market analysis for analytical and methodological reasons. Analytically, average costs indicate relative costs across jurisdictions. For example, Colby's (1990b) study reported results in terms of average transaction costs per acre foot at the state level. The second reason for the average transaction costs metric is methodological. Marginal transaction costs would require estimates of transaction costs per unit of water per transaction rather than transaction costs per unit of water per sub-basin. Transaction projects may take several years to complete; as such accurate data on transaction level costs are unlikely. McCann et al. (2005: 532) identify the trade-off between accuracy and precision, and the need to fit transaction costs metrics to the analytical purposes at hand, which in this case focused on a relative indication of transaction costs across jurisdictions and over time (sub-basin and annual averages) rather than across transactions and quantity (marginal).
5. The size of the budget in early years will also reflect the upfront costs of policy enactment to strengthen enabling conditions and adapt to unintended consequences. The overall size of the budget will reflect the transaction costs of getting to scale, that is, defined by the level of unmet environmental needs. It should be noted that determining environmental needs is an iterative and dynamic process.
6. See Chapter 2, citing Marshall (2013), who defines transaction costs as the costs of the resources to: 'define, establish, maintain, use and change institutions and organisations and define the problems that these institutions and organisations are intended to solve'.
7. For example, NPCC (2011) called for the CBWTP to justify its budget requests, noting: 'Cost monitoring is needed. Thirty six percent of the budget is to support QLEs [qualified local entities]. This is a big investment, and CBWTP should systematically evaluate how to keep acquisition and administration costs as low as possible. They could provide some analytical evidence of why this amount is needed to implement the project, because the National Fish and Wildlife Foundation (NFWF) as the lead entity should be able to assess the cost-effectiveness of the various approaches. This could be summarized in the annual meetings so that each QLE can learn from the experiences of other QLEs. The Consultant's evaluation report did not address the question of administrative efficiency or cost per acre-foot of leased or acquired water under different acquisition strategies. This could include a comparison of the annualized costs for a lease (with the accompanying multiple transaction costs) and outright permanent acquisitions (with the one-time accompanying transaction costs)'.
8. It must be noted that although the physical characteristics vary at multiple scales, the region shares defining features of its hydrology and hydrogeology, such as snowmelt-

driven hydrology with peak runoff in the spring and the natural low flow in the hydrograph. The presence or absence of losing reaches and reservoir storage alter the hydrogeologic character but in ways that are not expected to be determinative for performance trends.

9. The Willamette and Flathead sub-basins of Oregon and Montana, respectively, were no longer the focus of transactional activity within the CBWTP during the second period, while the levels in the Clark Fork and Umatilla were also limited during this second period (2008-10).
10. At this stage of implementation, although each unit of water recovered does not generate equal levels of ecological and economic benefits, the volume of unmet demand for environmental assets is high enough across the sub-basins selected for analysis that the initial acquisitions are at the high end of the demand curve. An assessment of marginal ecological benefits has not been developed to translate flow increases to differential levels of ecological impact.
11. The absence of comprehensive, systematic and high-resolution flow targets makes it impossible to use metrics based on water recovered as a proportion of estimated flow needs. As noted above, initial assessments of ecological impact are inconclusive due to a lack of established flow needs (which itself is a key part of the policy enactment and planning transaction costs) and the role of other limiting factors causally linked to ecological conditions.
12. Converting the cost per CFS to a cost per acre foot (reported for other water market analysis) requires assumptions about the duration of the contract. If the CFS reallocated instream were diverted for a standard irrigation season (180 days), one CFS would be approximately 360 acre feet.
13. Corresponding to the top right quadrant with high recovery and high water rights reform.
14. Corresponds to the top left quadrant of Figure 4.2.
15. Corresponding to the middle zone of Figure 4.2, including those such as the Salmon and John Day, which straddle the cross-over point between the top and bottom sections.
16. A term used by a former director of an implementing body in the Pacific Northwest.
17. Who incidentally claimed the highest proportion of his full-time equivalent (FTE) dedicated to instream flow transactions of all other watermasters in the Columbia Basin sub-basins in the state.

REFERENCES

ALLEN CONSULTING GROUP. 2006. Transaction costs of water markets and environmental policy instruments. Mebourne: Allen Consulting Group.

AYLWARD, B. and NEWTON, D. 2006. Long range water management in Central Oregon. *Water Report*, 29, 1–12.

ANDERSON, T.L. and LEAL, D.R. 2001. *Free Market Environmentalism*. New York: Palgrave Macmillan.

BASURTO, X. and OSTROM, E. 2009. The core challenges of moving beyond Garrett Hardin. *Journal of Natural Resources Policy Research*, 1, 255–9.

BJORNLUND, H. 2004. *What Impedes Water Markets*. Sydney: Moin & Associates.

BLACKFOOT CHALLENGE. 2008. The Blackfoot Challenge. http://blackfootchallenge.org/Articles/.

BROMLEY, D.W. 1982. Land and water problems: an institutional perspective. *American Journal of Agricultural Economics*, 64, 834–44.

BROMLEY, D.W. 1989. *Economic Interests and Institutions: The Conceptual Foundations of Public Policy*. Oxford: Basil Blackwell.

CAREY, J., et al. 2002. Transaction costs and trading behavior in an immature water market. *Environment and Development Economics*, 7(4), 769–95.

CHALLEN, R. 2000. *Institutions, Transaction Costs and Environmental Policy*. Cheltenham, UK and Northampton, MA, USA: Edward Elgar Publishing.

COASE, R.H. 1960. The problem of social cost. *Journal of Law and Economics*, 3, 1–44.

COGGAN, A., BUITELAAR, E., WHITTEN, S. and BENNETT, J. 2013. Factors that influence transaction costs in development offsets: who bears what and why? *Ecological Economics*, 88, 222–31.

COGGAN, A., WHITTEN, S.M. and BENNETT, J. 2010. Influences of transaction costs in environmental policy. *Ecological Economics*, 69, 1777–84.

COLBY, B.G. 1990a. Enhancing instream flow benefits in an era of water marketing. *Water Resources Research*, 26, 1113–20.

COLBY, B.G. 1990b. Transactions costs and efficiency in Western water allocation. *American Journal of Agricultural Economics*, 72, 1184–92.

COLE, D.H. 2002. *Pollution and Property: Comparing Ownership Institutions for Environmental Protection*. Cambridge: Cambridge University Press.

COMMONS, J.R. 1931. Institutional economics. *American Economic Review*, 21, 648–57.

DEMSETZ, H. 1967. Toward a theory of property rights. *Law and Economics*, 1, 341–53.

DESCHUTES RIVER CONSERVANCY (DRC). 2012. *Annual Report*. Bend, OR: Deschutes River Conservancy.

EASTER, W.K., ROSEGRANT, M.W. and DINAR, A. 1998. *Markets for Water: Potential and Performance*. Boston, MA: Kluwer Academic Publisher.

EGAN, T. 2011. *The Good Rain: Across Time and Terrain in the Pacific Northwest*. New York: Knopf Doubleday.

GARRICK, D. 2010. Explaining institutional diversity in emerging markets for environmental flows: a transaction costs approach to comparative policy evaluation in the Columbia Basin. University of Arizona: UMI Dissertation Publishing.

GARRIDO, S. 2011. Governing scarcity: water markets, equity and efficiency in pre-1950s eastern Spain. *International Journal of the Commons*, 5, 513–34.

GOVERNMENT ACCOUNTABILITY OFFICE (GAO). 2002. Columbia River Basin Salmon and Steelhead: federal agencies' recovery responsibilities, expenditures and actions. Washington, DC: GAO.

HARDNER, J. and GULLISON, R.E. 2007. *Independent External Evaluation of the Columbia Basin Water Transactions Program (2003–2006)*, Amherst, NH: Hardner & Gullison Consulting.

HAYEK, F.A. 1945. The use of knowledge in society. *American Economic Review*, 35, 519–30.

HEARNE, R.R. and EASTER, K.W. 1995. *Water Allocation and Water Markets: An Analysis of Gains-from-Trade in Chile*. Washington, DC: World Bank Publications.

HEIKKILA, T., E. SCHLAGER and DAVIS, M.W. 2011. The role of cross-scale institutional linkages in common pool resource management: assessing interstate river compacts. *Policy Studies Journal* 39(1), 121–45.

HOWITT, R.E. 1994. Empirical analysis of water market institutions: The 1991 California water market. *Resource and Energy Economics*, 16, 357–71.
HUBERT, G., GOLDEN, B. and MCCAULOU, S. 2009. Permanent environmental flow restoration through temporary transactions. International Conference on Implementing Environmental Water Allocations, Port Elizabeth, South Africa.
LOVRICH, N.P., SIEMANN, D., BROCK, J., BIRELEY, R.M., GAFFNEY, M.J., KENT, C. and HUCKABAY, J. 2004. Of water and trust: a review of the Washington Water Acquisition Program. Policy Consensus Center, prepared for the Washington State Department of Ecology and the Division of Governmental Studies and Services at Washington State University, Olympia, United States.
MACDONNELL, L.J. 1990. *The Water Transfer Process as a Management Option for Meeting Changing Water Demands*. Boulder, CO: Natural Resources Law Center, University of Colorado.
MARSHALL, G. 2005. *Economics for Collaborative Environmental Management: Regenerating the Commons*. London: Earthscan.
MARSHALL, G.R. 2013. Transaction costs, collective action and adaptation in managing complex social–ecological systems. *Ecological Economics*, 88, 185–94.
MCCANN, L., COLBY, B.G., EASTER, W.K., KASTERINE, A. and KUPERAN, K.V. 2005. Transaction cost measurement for evaluating environmental policies. *Ecological Economics*, 52, 527–42.
MCCANN, L. and EASTER, W.K. 2000. Estimates of public sector transaction costs in NRCS programs. *Journal of Agricultural and Applied Economics*, 32, 555–63.
MCCANN, L. and EASTER, W.K. 2004. A framework for estimating the transaction costs of alternative mechanisms for water exchange and allocation. *Water Resources Research*, 40. Available at: http://onlinelibrary.wiley.com/doi/10.1029/2003WR002830/full.
METTEPENNINGEN, E., BECKMAN, V. and EGGER, J. 2011. Public transaction costs of agri-environmental schemes and their determinants – analyzing stakeholders' involvement and perceptions. *Ecological Economics*, 70, 641–50.
MOORE, D., WILLEY, Z. and DIAMANT, A. 1995. Restoring Oregon's Deschutes River, developing partnerships and economic incentives to improve water quality and instream flows. New York: Environmental Defense Fund and Confederated Tribes of the Warm Springs Reservation of Oregon.
NEUMAN, J.C. 2004. The good, the bad, and the ugly: the first ten years of the Oregon Water Trust. *Nebrasksa Law Review*, 83, 432–84.
NEUMAN, J.C., SQUIER, A. and ACHTERMAN, G. 2006. Sometimes a great notion: Oregon's instream flow experiments. *Environmental Law*, 36, 1125–55.
NORTH, D.C. 1990. *Institutions, Institutional Change and Economic Performance*. Cambridge: Cambridge University Press.
NORTH, D.C. 1994. Economic performance through time. *American Economic Review*, 84, 359–68.
NORTHWEST POWER AND CONSERVATION COUNCIL (NPCC). 2005. *Deschutes Subbasin Plan*. Portland, OR: Columbia River Basin Fish and Wildlife Program.
NORTHWEST POWER AND CONSERVATION COUNCIL (NPCC). 2010. Final RME and artificial production categorical review report. In:

INDEPENDENT SCIENTIFIC REVIEW PANEL (ed.). Portland, OR: NPCC.
NORTHWEST POWER AND CONSERVATION COUNCIL (NPCC). 2011. Expenditures report: Columbia River Basin Fish and Wildlife Program. In: *Annual Report to the Northwest Governors*. Portland, OR: NPCC.
OSTROM, E. 1990. *Governing the Commons: The Evolution of Institutions for Collective Action*. Cambridge: Cambridge University Press.
OSTROM, E. 2009. A general framework for analyzing sustainability of social-ecological systems. *Science*, 325, 419–22.
PILZ, R.D. 2006. At the confluence: Oregon's instream water rights in theory and practice. *Environmental Law Journal*, 36, 1383–420.
POSNER, R. 1986. *Economic Analysis of Law*. Boston, MA: Little, Brown.
RUML, C.C. 2005. The Coase theorem and western U.S. appropriative rights. *Natural Resources Journal*, 45, 169–200.
SCHLAGER, E. and BLOMQUIST, W.A. 2008. *Embracing Watershed Politics*. Boulder, CO: University Press of Colorado.
SIEMANN, D. and MARTIN, S. 2007. Managing many waters: an assessment of capacities for implementing water and fish improvements in the Walla Walla Basin. Pullman, WA: William D. Ruckelshaus Center.
SINGLETON, S.G. 1998. *Constructing Cooperation: The Evolution of Institutions of Comanagement*. University of Michigan Press.
WEBER, E.P., LOVRICH, N.P. and GAFFNEY, M. 2005. Collaboration, enforcement, and endangered species: a framework for assessing collaborative problem-solving capacity. *Society and Natural Resources*, 18, 677–98.
WILLIAMSON, O.E. 1998. Transaction cost economics: how it works; where it is headed. *De economist*, 146, 23–58.

APPENDIX: CODING EXCLUSION, TRANSFERABILITY AND ADMINISTRATIVE CAPACITY OF WATER RIGHTS REFORM AT THE SUB-BASIN LEVEL

Table 4A.1 Exclusion (level)

	Coding question
Closure	Cap (aggregate boundary): Is the sub-basin closed to further extraction? How?
Instream flow	Have environmental flow requirements been defined and promulgated by rule or another formal collective choice process?
Adjudication	Water rights (boundary between users): Are the extent, validity, and priority of the water right specified in a court decree of adjudication proceedings?
Forfeiture: beneficial use	Do operational rules uphold restrictions against non-use?
Transboundary status	Is the sub-basin bisected by state political or tribal sovereign boundaries?
Mitigation	Is mitigation required for groundwater pumping, including permit-exempt wells, in closed basins or within hydrologically connected zones with surface water rights vulnerable to capture?

Note: Attributes were coded based on the presence/absence of each sub-dimension as well as the type, scope, and completeness of exclusion mechanisms enacted.

Table 4A.2 Transferability

	Coding question
Contract type	Are irrigation efficiency savings, or other positive sum contract types, permitted?
Contract duration	Are permanent transactions authorized?
Burden of proof: beneficial use	Is a tentative determination of validity and extent required to establish proof of historic use required for long-term environmental water transactions?
Clarity / ease of administrative procedure: injury analysis	Do instream-specific rules exist to interpret and apply statutory authority for leases and transfers, including standards for administrative review?
Banking institution	Do banking institutions or rotational pools exist to coordinate leasing or transfers?
Legal incentive: protection against forfeiture	Do instream leasing/ donations afford protection against forfeiture?

Table 4A.3 Administration

	Coding question
Watermaster	Does a local watermaster exist? If so, how many and at what level of spatial scale, i.e., what portion of the watershed? Who appoints or elects the watermaster and how?
Measurement	Is measurement required for large diversions? If so, under which conditions?
Monitoring	Are water supply (streamflows) and use (diversions) actively monitored to guide regulation activities? To what degree and level of coercion (for example, voluntary or mandatory)?
Regulation	Does a watermaster regulate water rights according to the prior appropriation doctrine? Is regulation activity complaint-driven or trigger-based?
Conflict resolution	Does a low-cost, informal option exist for conflict resolution at the local level?
Watershed	Is statutory or regulatory authority for measurement, monitoring, or enforcement explicitly developed at a field office within administrative units defined at the watershed scale?

5. Maturing water markets and public goods in the Murray–Darling Basin: scaling up water trading and transboundary governance

> Afterwards he ceased to be a miner . . . but believing that he had an exclusive right to the enjoyment of the water over which he had control for a long period, he did not hesitate to sell it; and in this manner, by slow growth as it were, the claimholder was transformed into an owner of water.
> (R.B. Smyth, 1869 (Powell, 1989): 50, capturing the long history of water rights trading in Victoria)

> [T]rusts should be induced to amalgamate, so as to embrace territories remodeled on a natural basis, each, as far as possible, including the entire supply from one watershed, with one compact area of distribution. These watershed trusts should have control, under proper regulations, of the entire distribution of water within their territories. (Deakin, 1885 (Powell, 1989): 113)

> [M]anagement of Australia's environment by way of its catchment systems should be strengthened. . . [T]his approach will be more cost effective and will reliably and efficiently attain the outcomes needed.
> (Australia Parliament House of Representatives Standing Committee on Environment Heritage, 2001: 38)

INTRODUCTION

The Murray–Darling spans more than 1 million square kilometres and six major jurisdictions – four states, one territory, the Commonwealth government – and a number of irrigation regions, cities and indigenous communities. Outflows from the Murray Mouth are 40 per cent of their pre-development levels due to the combined effects of dams, diversions and drought. The upstream–downstream interdependencies associated with river basin closure have triggered efforts to 'scale up' water planning and allocation institutions to achieve sustainable outcomes. These factors prompted efforts to craft integrated water markets that enhance adaptive efficiency, equity and sustainability at multiple nested scales.

In this chapter, polycentric governance theory provides a lens to analyze the co-evolution of water markets and river basin governance arrangements across three linked components of reform: water rights reform, capping and environmental water recovery. I frame the challenge of scaling up water allocation reform as an institutional collective action dilemma and review the history of integrative river basin governance. Transaction costs can both enable and constrain integration. Coordination efforts, consensus-building and other forms of collaboration and conflict resolution can build legitimacy and trust needed to establish institutional frameworks that reduce transaction costs over the long term. However, integration is costly, takes time and is limited to situations where benefits justify the transaction costs involved. In the presence of transaction costs and limited compliance, fragmentation persists and integration mechanisms at the basin scale have struggled to balance subsidiarity (decisions at the lowest level possible) with cross-scale governance capacity to coordinate trade-offs and public goods provision across jurisdictional boundaries. I assess trends and types of integration mechanisms, which differ by their authority and scope (working groups, basin organizations, intergovernmental agreements). A portfolio of formal and informal mechanisms has developed rather than a unified and comprehensive basin-wide mechanism. Scaling up has been marked by a dynamic tension between water markets, which require efforts to reduce transaction costs, and basin governance arrangements, which impose transaction costs to coordinate the legitimate interests and capacities of different central and local governments and stakeholders.

The evolution of polycentric governance arrangements and coordination mechanisms in the Murray–Darling can be understood as a swing of a pendulum between the extreme points of consensus-based interstate arrangements to a strong federal authority which triggered resistance from both states and local users. The contemporary position lies between these points and reflects a middle path marked by subsidiarity, institutional diversity and a portfolio of formal and informal institutional mechanisms for basin-wide coordination.

THE CHALLENGE OF SCALING UP: FROM THE COLUMBIA TO THE MURRAY–DARLING

The Murray–Darling River failed to reach the sea in 1981 and 2002 – evidence of river basin closure due to the convergence of development, demand and droughts. These events revealed structural imbalances between supply and demand, and prompted a transition to tradable water rights and river basin governance to establish sustainable water diversion limits. Water trade is nothing new, however, as illustrated by Smyth's observations of 1860s Victoria, above. Nor are efforts to promote hydrological boundaries as the ideal units for water management, which date (at least) from the Royal Commission on Water Supply in the 1880s and paralleled John Wesley Powell's arguments in the Western US during this period. However, efforts to marry water markets and river basin governance are more novel, recent and complex, involving different actors, incentives and levels of governance (Bauer, 2004).

Integrated water markets combine water rights reform and basin governance through a set of water policy, planning and legal reforms at multiple scales. Grafton et al. (2011) define an 'integrated water market' in terms of 19 criteria across three dimensions: institutional foundations, efficiency and environmental sustainability. The institutional foundations include not only the conventional prerequisites for markets – administrative capacity, legal clarity over property rights, recognition of the public trust, and so on – but also transboundary governance arrangements to scale up markets through 'horizontal' and 'vertical' coordination mechanisms. Horizontal coordination is concerned with interactions within each level of governance, such as the coordination of states sharing a river, or water users and user groups with a stake in a river or catchment (for example irrigation, cities, environment) (Berkes, 2002). Vertical coordination mechanisms (Young, 2002) structure interactions between different tiers of governance, such as those between state and federal governments and those between resource users and states. River basin governance and interstate agreements create an additional governance layer. Integrated water markets also require adaptive institutions, defined as the capacity to cope with crises, assimilate new information and adjust water management as social values evolve (Grafton et al., 2011).

In the Columbia Basin, the path toward integrated water markets and adaptive institutions has been a 50-year process and remains elusive, as described by Chapter 4. The reform process began as early as the 1950s when fisheries began to suffer from the cumulative impacts of water diversions (Neuman et al., 2006; Lichatowich, 2001). Since the 1970s, salmon recovery efforts have established watersheds as the basis for freshwater conservation and watershed planning. Administrative units for water rights in Oregon and Washington are defined within the same boundaries (sub-basins) underpinning salmon recovery efforts and associated watershed governance arrangements. These policy reforms have established the enabling conditions for integrated water markets at the sub-basin level, including de facto caps on diversions through minimum flows, and water trading rules that either account for environmental impacts (requiring mitigation to offset new groundwater pumping) or contribute directly to environmental recovery (acquisition of water for the environment). Yet, trading activity is limited, and many sub-basins remain overallocated and dewatered in late summer. New demands for residential development are often met through unsustainable groundwater pumping, exacerbating the closure of tributaries.

It has been ten years since the creation of the Columbia Basin Water Transactions Program, and success in the basin remains highly localized at (or below) sub-basin scales, not the basin as a whole. Fragmented progress can be attributed partly to the fact that scarcity and competition remain localized and episodic (that is, during late summer and periodic droughts) rather than systemic and chronic. Individual tributaries are closed or closing, but not the basin as a whole when measured in terms of outflows to the sea. Demand for environmental flow restoration has provided a primary catalyst for markets to emerge, rather than intensified competition between irrigators and cities. As a result, trading between consumptive users and uses (for example from agriculture to cities) has been patchy and limited by the ad hoc nature of legal, administrative and environmental limits on water consumption. Important exceptions exist in the Deschutes, Kittitas Valley (Yakima) and parts of Montana where residential development in rural communities has created pressure on groundwater stocks and created the political will and regulatory framework for mitigation efforts. Even in these regions, trade across jurisdictions remains limited, which stifles recovery in tributaries shared by multiple large irrigation districts or across state borders. For example, the Walla Walla of Northeast Oregon and Southwest Washington is governed by two different state jurisdictions which makes water acquired and conserved for fish in upstream Oregon vulnerable to depletion as it crosses into Washington (Siemann and Martin, 2007). Even when a 'winning coalition' of irrigators, cities and environmental interests coalesces around proposed institutional changes, sustainable and adaptive water allocation reforms have languished. Nevertheless, the Deschutes and Salmon rivers demonstrate what can happen when river basin arrangements and markets become aligned through multilevel collective action. Entrepreneurial place-based initiatives have the potential to restore environmental flows and achieve sustainable water allocation reforms when state- and community-based planning efforts impose limits on water extractions, and irrigators respond to new incentives within a market-based allocation framework.

SCALING UP AS AN INSTITUTIONAL COLLECTIVE ACTION DILEMMA

Localized progress in the Columbia begs the question: whether and how to 'scale up' water allocation reform in closed basins, as upstream and downstream users and jurisdictions become increasingly interdependent? The local success stories of the Columbia demonstrate that market-based reallocation and environmental water recovery can provide an effective and politically attractive institutional alternative to court and administrative methods. But can local water rights and catchment management reforms contribute to sustainable and adaptive river basin governance arrangements across multiple jurisdictions and competing public goods? And what types of polycentric governance arrangements enable or constrain this potential? These are the questions of institutional design and change addressed in this chapter. Theories of local public economies and institutional collective action provide a lens for understanding this institutional evolution and considering current and future trajectories of change. Both theoretical perspectives acknowledge transaction costs as a limiting factor on integration, resulting in polycentric governance arrangements in which fragmentation is balanced by coordination[1] (Ostrom et al. 1961, Pahl-Wostl et al., 2012). They also distinguish between different types of transaction costs to highlight that transaction costs minimization is not always desirable; instead there is value in incurring transaction costs to build institutions and enable integration when the collaboration risk and public goods justify the investment in the decision-making, administration and commitment costs (Schlager and Blomquist, 2008).

The scaling-up challenge presents dilemmas of 'institutional collective action'. Feiock (2013) refers to institutional collective action (ICA) dilemmas[2] as situations where distinct governmental jurisdictions become interdependent and must cooperate to fulfil their goals and overlapping functions (see below). Dilemmas arise due to incoordination, unfair division and defection, which arise at the intersection of water markets and river basin governance in the form of upstream–downstream conflicts over water apportionment, rent-seeking by agencies which may block changes to maintain their budgets, and competition between states and federal governments over decision-making authority and fiscal arrangements. In response to ICA dilemmas, efforts to achieve integrated water markets and river basin governance arrangements strive to manage the basin as a coherent 'public economy', where a public economy refers to the organization of public goods provision and production, and acknowledges that fragmentation.

Polycentricity is hypothesized to enhance adaptive efficiency by balancing efficiency with other policy objectives associated with the provision of public goods, such as representation, accountability and and legitimacy. Polycentricity can foster efficiency, accountability and resiliency in the provision of public goods through competitive rivalry and experimentation, while redundancy reduces the potential for local failures to lead to cascading impacts and even system-wide failures. Conceptualizing river basins as public economies allows for a rigorous assessment of decision-making and the delivery of water-related public goods. It can be used to identify who provides and produces, and to identify the tensions between integration (in capping water diversions and the management of third party effects) and decentralization (private trading behavior and local management by irrigation organizations or catchment management authorities) in water allocation reform. As such, the concept of a public economy provides a lens to understand the complex arrangements for public goods provision and production in river basins and trace how this has shaped the evolution and performance of water markets and sustainable water allocation reform. This theoretical perspective posits that integration is not necessarily a good thing when it depends on strong centralization and comes at the expense of institutional diversity by crowding out local and informal institutions. Instead, 'scaling up' is expected to rely on institutional diversity and principles of sophisticated subsidiarity: ensuring decision-making at the lowest level with authority and capacity to do so, but no lower (Marshall, 2008). The resulting institutional landscape is expected to be polycentric with performance influenced by the distribution of authority and mechanisms for vertical and horizontal coordination (Pahl-Wostl and Knieper, 2014).

The Murray–Darling Basin is the last destination in the tour of three river basins addressed in this book and represents an ideal laboratory to investigate these patterns of institutional change. There are 23 major valleys across parts of four states and one capital territory comprising the Murray–Darling Basin (Murray–Darling Basin Authority, 2010). At least 48 'social catchments' have been identified by efforts to determine communities with a 'distinct identity and coherence, which were likely to respond differently [to basin planning efforts]' (EBC et al., 2011: 10). The Southern Connected Murray (SCM) refers to the rivers in the Southern Murray with a nearly permanent hydrologic connection. This hydrologic and infrastructure connectivity enables water to move from the upstream catchments to be managed and delivered downstream. In the past two decades, water trading, basin governance arrangements and environmental water recovery programs have been established at the local, state, basin and Commonwealth levels. Given the hydrological and operational connectivity of the SCM, political boundaries are a primary impediment to basin integration. The Murray–Darling Basin has undergone two decades of reform and billions of dollars in federal and state government expenditures to establish water markets and build river basin governance institutions (Lee and Ancev, 2009).

Scarcity, drought and competition prompted the 'unbundling' of water rights – separating water and land property rights – in the late 1980s. Efforts to harmonize water rights systems, devise and adjust sustainable diversion limits, and recover water for the environment have led to a progressive scaling-up of reform from state to basin-wide and national coordination, requiring a balancing of local capacity and decentralized decision-making by irrigators and irrigation communities with new public interests and intergovernmental authorities. Progress has not been linear and integration remains partial.

The reform process has foundered on the question of which management level is best suited for which tasks, and how these roles and responsibilities should be coordinated (Marshall et al., 2013; Marshall, 2008). Examining the recent history of reform through the perspective of local public economies will also shed light on institutional design issues for lingering or looming intergovernmental dilemmas: interstate trade, groundwater trade, environmental water management and harmonization of water resource plans under new federally devised sustainable diversion limits (SDLs). The evolution and design of polycentric governance arrangements in the presence of substantial transaction costs is therefore a major concern for public policy. Lessons from past and ongoing reforms can be used to navigate future trajectories and build adaptive capacity to cope with complex trade-offs and interdependencies in closed rivers.

The rest of the chapter defines the challenge of scaling up, and it reviews the evolution of integrated river basin governance. I draw on conceptual and analytical perspectives at the intersection of theories of local and public economies and institutional collective action to elaborate a typology of coordination mechanisms for basin governance institutions. The different forms of scaling up are examined across three phases of institutional change at the intersection of water markets and river basin governance. This historical analysis is used to understand the relationship between transaction costs and integrated water markets and to explain why the quest for centralized and comprehensive basin-wide reform remains elusive. In so doing, I argue that water allocation reforms should design for diversity by balancing subsidiarity (local capacity in informal institutions) and complementarity (a portfolio of formal and informal coordination institutions).

Table 5.1 Scaling up: institutional change and coordination

Scaling up (criterion)	Institutional change	Evolution of integration mechanisms (basin-wide)
Water trade (economic efficiency)	Unbundle land and water rights	• Working Group on Property Rights • Intergovernmental Agreement (National Water Initiative)
	Establish trading rules	• Working Group for Pilot Interstate Trade • Basin Plan Ch 12 and Australia Competition and Consumer Commission
Cap (sustainability)	Audit and monitor water use	• Intergovernmental Agreement (MDB Agreement)
	Establish diversion limits	• Council of Australian Governments water reform • Commonwealth Water Act • Murray–Darling Basin Authority • Basin Plan 2012 (statutory)
Environmental recovery (adaptability)	Identify environmental needs	• The Living Murray (IGA) → 2008 Commonwealth Water Act Amendment
	Acquire water for the environment	• Commonwealth Environmental Water Office
	Deliver water for the environment	• Basin Plan 2012 (statutory)

WHAT DOES SCALING UP MEAN?

Scaling up has at least three meanings in cap-and-trade allocation reform (Table 5.1). First, water markets scale up when transactions enhance economic efficiency[3] and lead to gains from trade, following the logic of water markets outlined at the outset of Chapter 2. Reallocation enhances economic welfare when the marginal value of production varies across uses and users, and water users are free to exchange water rights. Economic models have assessed the potential gains from water trade in a range of settings based on assumptions about cropping patterns and commodity prices (for example, Pujol et al., 2006; Garrido, 1998; Qureshi et al., 2009). However, these potential gains from trade have failed to materialize in water-stressed regions as diverse as Italy, Spain, the Western US and China. Transaction costs may impede otherwise welfare-enhancing trading behavior by preventing water reallocation to the uses and users with the highest economic value. Scaling up water trade in this context requires investment in the transition costs to set up the market, create tradable water rights and build administrative capacity to reduce transaction costs and manage social and environmental impacts of trade. In the Murray–Darling, efforts to strengthen water markets have been described in conventional neoclassical economic terms (Allen Consulting Group, 2006: 3):

> gains from trade need to be viewed in the context of the public and private transaction costs associated with establishing and maintaining water markets . . . the relevant measure of net efficiency improvement is whether the gains to society from trade outweigh the total 'efficient' transaction costs incurred by government and market participants [where 'efficient costs refer to the least cost combination of inputs for undertaking a trade, subject to meeting environmental and social objectives'].

Upon closer reading, the environmental and social objectives are situated outside of the market, and gains from market-based reallocation presuppose that a wider institutional framework is in place to provide the public goods associated with water. The second form of scaling up relates to sustainability and equity outcomes: diversion limits and trading rules to account for social and environmental impacts of water trade. Environmental and social objectives are the focus of the capping process and the associated regulatory framework to limit negative consequences of water trade (National Water Commission, 2011a, 2011b). These impacts are increasingly considered at the basin scale due to river basin closure. The second form of scaling up requires a nested set of sustainable diversion limits and regulatory systems.

Water markets are likely to emerge in response to scarcity and basin closure. By definition, such systems are typically overallocated, with insufficient environmental flows to sustain freshwater ecosystems and the goods and services they deliver for people and nature. The third form of scaling up is needed to address such overallocation. In overallocated regions – in which the proposed cap is below existing diversion levels – the capping process also entails a transition to new diversion limits by recovering water from existing users, akin to the experience in tributaries of the Columbia. Scaling up environmental water recovery requires adaptability to adjust both water rights systems (to enable environmental entitlements) and diversion limits (to ensure sustainable levels of extraction, often by acquiring water entitlements for the environment). Environmental water recovery in the Murray–Darling comprises a landscape-level challenge across a territory twice the size of France. The political economy of environmental water recovery is contentious. Scaling up pits irrigators against other stakeholders due to the local, concentrated and immediate costs for irrigators and the diffuse, delayed and distributed benefits for a range of constituencies.

Scaling up water rights reform, sustainable diversion limits and environmental water recovery has continuously redefined the boundary and interactions between decentralized behavior (trading) and collective choices (capping, regulation of third party effects, environmental water recovery). The three forms of scaling up are interwoven, even though the capping process is taken as largely exogenous and *ex ante*. The need to set up and periodically modify the institutional arrangements governing water markets contradicts the proposition that water markets are a 'self-maintaining allocation mechanism', an argument that prevailed during the free market environmentalism era but has been persistently debunked or challenged (Challen, 2000; Bell and Quiggin, 2008; Connell et al., 2005). A self-regulating market for allocating scarce water resources implies a firewall that protects the market from political allocation decisions and government intervention (Anderson and Leal, 2001; Zetland, 2011). The reform experience in the Murray–Darling reveals the challenge of upholding this premise. Bell and Quiggin cite two reasons for the retreat of the market logic, at least in its purest form: (1) setting up water rights systems entails high transaction costs and requires government capacity to provide clear flows of information, reporting and administration; and (2) markets will not deliver adequate protection for the environment without a strong government role. On top of these reasons, the political economy matters because scaling up creates winners and losers (states, irrigators and so on). The vested interests of irrigators are deeply entrenched and give rise to the associated political pressure exerted by irrigation organizations and state governments to restrict water from leaving their jurisdictions (see Chapter 3 in this volume). These vested interests led to voluntary buybacks instead of compulsory reallocation. Even voluntary buybacks have encountered stubborn resistance, however, due to perceived broader impacts of environmental water recovery on irrigation districts and regional communities. Viewing the three forms of scaling up together and as part of an adaptive process, however, provides a more complete picture of transaction costs, institutional change and adaptive efficiency.

THE MURRAY–DARLING AS A LABORATORY

Water markets in the Murray–Darling are viewed with envy internationally. So are the region's basin governance arrangements. Australia's experience has been touted for its insights relating to scaling up adaptive governance at the intersection of water markets and basin governance institutions. Of the five countries and regions (Australia, the US, Chile, China, South Africa) compared by Grafton et al. in their study of integrated water markets, 'at least at the present time, Australia has the most adaptable institutions, although, as elsewhere, political tensions affect the pace and degree of reform implementation' (Grafton et al., 2011: 228). Over the past two decades, and particularly since 2004, trading activity in Southeast Australia has increased and become more embedded within river basin governance arrangements to coordinate planning and allocation within and across river valleys, irrigation systems and state jurisdictions.

Consider the first type of scaling up: water trade to achieve gains from trade. Water trade in the Murray–Darling now accounts for the vast majority of water transactions in Australia, representing more than 80 per cent of such transactions (National Water Commission, 2011b). Trading activity includes the buying and selling of water entitlements (permanent access to a share of the water available) and allocations (temporary use of the volume of water assigned to a given class of entitlements in a given year). Entitlement and allocation trade is analogous to permanent and temporary water trade, respectively, in the Western US. Like the Western US, temporary transactions predominate in part due to the comparably low transaction costs and processing times, but also due to the sophisticated asset management strategies of irrigators (Loch et al., 2013). Water allocation trade had increased to as much as 38 per cent of total surface water use in 2011–12 after rapid growth during the late stages of the Millennium Drought (1997–2010), which continued after the drought broke in 2010 (Grafton and Horne, 2014). This trend continued into late 2013. August, September and November 2013 are among the most active periods of temporary water trade (in terms of volume) on record; permanent trade is a small but important part of the market, particularly for environmental acquisitions. Irrigators have become strategic and sophisticated by relying principally on temporary trading in their risk management (Waterfind, 2013).

Next, consider the second and third forms of scaling up: establishing a cap and recovering water for the environment. In the Murray–Darling, demands on water resources have multiplied over the past century, posing complex and contentious trade-offs, not only between competing private uses and between public and private uses, but also between different public values. Integrated river basin governance has been upheld as the solution to these challenges, acknowledging the tension between fragmentation and coordination. The tension between local control and basin or national coordination has a long history, echoing Clark's reflections about the prospect that a single body could be necessary to govern the River Murray (Clark, 1971). 'One view is that successful river management is impossible unless there is one body which is the sole repository of administrative and regulatory power in relation to the Murray' (Clark, 1971: 251).

INTEGRATED RIVER BASIN MANAGEMENT IN THE MURRAY–DARLING: A BRIEF HISTORY

The statutory 2012 Basin Plan represents an important milestone in the quest for integrated river basin management. However, the architecture of interstate water governance in the Murray–Darling has been a subject of intense debate since well before federation in 1901 (Clark, 1971; Webster and Williams, 2012). Scaling up has been a persistent challenge since that time. Musgrave (2008) categorizes the history of water resource development into three phases: territorial legislation following European settlement in 1788; government-sponsored irrigation development in the late 1800s and particularly after federation in 1901; and a reform phase under way since the 1980s to manage drought, shifting needs and environmental values. Connell (2007) has also divided the history of institutional development and basin-wide integration into three phases, demarcated by the River Murray Waters Agreement (1915–92), the Murray–Darling Basin Initiative (1992–2007), and the Commonwealth Water Act (2007–present); see also (Marshall et al., 2013). Table 5.2 lists milestones in this history.

The Australian Constitution (section 100) is the foundation of this institutional evolution. It formally divides powers between federal and state governments for water management, reserving the 'right of a state or of the residents therein to the reasonable use of the waters of rivers for conservation or irrigation' (Clark, 1971; Webster and Williams, 2012). These constitutional provisions vested allocation authority at the state level. Other constitutional provisions (section 51, 98) provided scope for national and interstate cooperation, particularly those pertaining to trade and navigation. The 1902 Interstate Royal Commission on the River Murray set the path toward interstate water sharing after vigorous debates and political manoeuvres by the new state governments, culminating with the 1914–15 River Murray Waters Agreement. The agreement established water-sharing arrangements, provided for joint construction and maintenance of infrastructure, and led to the creation of the River Murray Waters Commission in 1917, which coordinated interstate river management until the mid-1980s (see Chapter 3 for more on the river basin trajectory).

Mounting environmental challenges since the 1970s, coupled with fiscal problems in recovering the costs of irrigation systems, created the impetus for reform by the 1980s (Blomquist et al., 2005; Harris, 2011, 2007). States

Table 5.2 Milestones in integrated river basin governance

1863	Inter-colonial conference to enhance navigability for commerce and trade
1895–1903	Federation Drought
1901	Constitution of Australia creates Australian Federation, 1 January 1901
1901	Constitutional powers related to water allocation reserved for states
1902	State representatives meet in Corowa, New South Wales to resolve disputes over River Murray
1914	River Murray Waters Agreement
	Divides water between states as shares and forms an interstate commission
1917	River Murray Commission established
1936	Hume Dam completed as a joint enterprise of New South Wales and Victoria
Post-World Wars	State-run schemes for soldier settlement in irrigation schemes
1985–86	Murray–Darling Ministerial Council formed
1992	Murray–Darling Basin Agreement
	Murray–Darling Basin Commission established to replace the River Murray Commission; water sharing arrangements basin-wide (extended to Queensland and Australian Capital Territory)
1994	Council of Australian Governments water reform package
1995–97	Interim cap on extractions
	Murray–Darling Basin Commission conducts audit of water use and states agree to impose interim cap based on water usage under 1993–1994 development
1997–2009	Millennium Drought
2003	Intergovernmental Agreement to establish The Living Murray
	Interstate river restoration program established
2004	Intergovernmental Agreement to establish the National Water Initiative
	Overhauls management, measurement, planning, pricing and trading
2007	National Water Act
	Requires Basin Plan to limit diversions in sub-basins
2008	Murray–Darling Basin Authority established
	Responsible for planning 'integrated management' across the Basin
2012	Basin Plan enacted as law
	'Coordinated sustainable approach to water use'

were operating independently and were centralized internally with capacity concentrated in capital cities far removed from irrigation regions. The relationship of the central and local governments presented a challenge for institutional collective action: within states, between the states, and between the states and the Commonwealth government (Blomquist et al., 2005).

The dredging of the Murray Mouth in 1981 signalled the heightened interdependency of the Murray as a closing system, requiring new modes of integration to address the institutional collective action challenges of water sharing and provision of public goods tied to sustainability and ecosystem health. The Murray–Darling Basin Initiative of the mid-1980s promoted integration and cross-jurisdictional cooperation well before the Dublin Principles established integrated water resource management as a global paradigm in 1992. Connell (2011a) identifies eight subsequent major attempts at integrated river basin governance:

- Salinity and Drainage Strategy 1989–90.
- Natural Resource Management Strategy 1990.
- Council of Australian Governments Water Reform Package 1994.
- Cap of 1995–97.
- Integrated Catchment Management Strategy of 2000.
- The Living Murray First Step 2003–04.
- National Water Initiative 2004.
- Water Act/Basin Plan 2007+.

The Murray–Darling Basin Agreement and ensuing intergovernmental agreements in 1994 and 2004 balanced local control (including cost recovery by privatized irrigation organizations) with basin-wide coordination. The Murray–Darling Basin Commission and Ministerial Council played an essential coordination role in this period, comprising representatives from the Commonwealth and state ministries for environment, land and water. States have reached agreements on interstate challenges tied to natural resource management, salinity management and environmental water needs. As a rule, these interstate agreements were noteworthy for being 'low in compliance and high in transaction costs . . . to protect state interests' (Connell, 2011a: 328). Milestones include the 1994 Coalition of Australian Governments water reform framework, which paved the way for an interim cap on diversions in 1995/7 and identified the environment as a legitimate water user. By the turn of the century, the Integrated Catchment Management Strategy, the National Salinity Action Plan and The Living Murray environmental water recovery initiative were arguably high-water marks for integrated basin governance. The National Water Initiative in 2004 attempted to harmonize statutory water planning and water rights across states. Interstate water management issues required political negotiation under a consensus decision rule, which allowed individual states to block change, shirk responsibilities or adopt inconsistent approaches to capping. The interim cap in the mid-1990s was a prime example of this coordination challenge. New South Wales insisted on recognizing sleeper and dozer licenses, while the establishment of a cap in Queensland and the Australian Capital Territory lagged for years before coming into force as part of an amendment to Schedule F of the Murray–Darling Basin Agreement in 2008.

The 2007 Water Act represented a fundamental shift in the balance between state and federal authority, and was a major attempt at comprehensive water management policy; in part as a result of frustration with the consensus decision rule and the inability of the states to cooperate on environmental water recovery during a decadal drought (Connell, 2007, 2011b, 2011a; Crase et al., 2011). The Act created two federal authorities – the Murray–Darling Basin Authority and the Commonwealth Environmental Water Office – with control over budgets and several operational policies allowing unilateral decisions about water planning and environmental water use, respectively. The expanded federal role under the Act invoked new sources of constitutional authority. The Act derives its authority from multiple constitutional sources, including state referral of powers to the Commonwealth under the 1992 Murray–Darling Basin Agreement, and the external affairs clause; particularly related to international obligations for biodiversity conservation (Fisher, 2011; Skinner and Langford, 2013).

The Act called for the development of a Basin Plan to define sustainable water diversion levels and establish environmental water requirements. The Basin Plan was adopted into statute in 2012 with all states and territories acceding to an intergovernmental agreement to implement the plan as of 2014. The Plan comprised a pyrrhic victory after a contentious public engagement process and interstate negotiation, raising the level of mistrust and hence the transaction costs of future cooperation and integration (Crase et al., 2013). The consultation process for the Basin Plan involved substantial transaction costs, particularly after the release of the guide heightened mistrust. The subsequent phase of consultation during the Draft Basin Plan entailed 175 meetings and 19 000 participants, a stark indication of the costs incurred to secure legitimacy and integration (Crase et al., 2013). The process triggered political resistance from irrigators and states, and spurred a populist movement founded on the concept of localism. Ongoing controversies over scaling up and shifting federal–state–local roles in water allocation and environmental water management need to be considered against this shifting institutional and political backdrop of water federalism. Australia's water history provides important context about who plans, implements and maintains, as well as how costs and risks are shared over time; states retain their constitutional authority for allocation and planning, while federal financing has played a key role in securing Commonwealth goals. At root, these tensions reflect a long history of state-run irrigation development and path dependency, which form stubborn impediments to integrated river basin management at a time of stress and rapid change.

FINDING ORDER AMIDST CHAOS: UNDERSTANDING WHY FRAGMENTATION PERSISTS

Interstate cooperation is the Achilles heel of integration. Integrated river basin governance arrangements depend on coordinated decisions by independent state governments. Intergovernmental agreements and policy statements enshrined integrated basin management as an overarching objective, although implementation lagged as a consequence of interstate politics (Connell, 2007, 2011b, 2011a; Blomquist et al., 2005; Connell and Grafton, 2011). Interstate coordination was based on a consensus decision rule among state governments since the Murray–Darling Basin Initiative of the mid-1980s until the 2007 Water Act conferred new central authority in a reconstituted basin organization at the Commonwealth level: the Murray–Darling Basin Authority. Notwithstanding these shifting central–local relations, integrated river basin management remains elusive, as illustrated by stipulations of the Council of Australian Governments (COAG) reforms, Parliamentary Committee on Integrated Catchment Management and the more recent concentration of federal authority since the 2007 Act.

The 1994 COAG reforms called for: 'administrative arrangements and decision-making processes to ensure an integrated approach to natural resource management . . . an integrated catchment management approach to water resource management and . . . arrangements to consult with the representatives of local government and the wider community in individual catchments'.

A 2000 parliamentary committee investigated 'the role of different levels of government, the private sector and the community in the management of catchment areas' leading to a proposal for a nationally coordinated system of catchment management with clear assignment of tasks across levels (Australian Parliament House of Representatives Standing Committee on the Environment and Heritage, 2000). The resulting integrated catchment management policy statement presented a comprehensive argument for integrated water management involving local capacity and national coordination.

A government review of the use of market mechanisms for environmental water recovery acknowledged the lingering challenges of fragmentation (Productivity Commission, 2010): 'Governance arrangements for the recovery and management of water for the environment are fragmented. Greater coordination of water recovery and environmental watering by Basin jurisdictions is required'.

Fragmentation persists despite these centralizing tendencies because of lingering basin-wide coordination challenges. Why? Connell (2007) notes that efficiency is not always the principal objective in efforts to scale up to basin-wide governance arrangements for water planning and allocation. Writing about The Living Murray program established in 2004 – the first river-wide environmental water recovery initiative for the Southern Connected Murray River – Connell notes that 'satisfying state demand to maintain . . . autonomy took precedence over operational efficiency' (Connell, 2007: 168). For basin-scale goals, legitimacy and representation of diverse interests often trumps efficiency in the narrow neoclassical sense. In The Living Murray experience, efforts to improve legitimacy and accountability included the use of state-by-state funding accounts, monitoring efforts and registries with joint oversight rather than a consolidated version coordinated centrally.

The proliferation and persistence of jurisdictional fragmentation arise in the pursuit of multiple, often competing, public goods; each with varying economies of scale, politics and jurisdictional complexity in their provision and production. Institutional collective action dilemmas stem from the need to coordinate across jurisdictions for public goods that span boundaries (for example, basin-wide sustainable diversion limits and meeting environmental water requirements).

The challenge of scaling up sustainable water allocation reform at the intersection of water markets and river basin governance is therefore a prime example of an institutional collective action dilemma: authority for public goods provision and capacity for their production are divided across multiple jurisdictions with only partial control over the evolution and outcomes of reform.[4] Solving or mitigating institutional collective action dilemmas can be viewed in terms of the basic political economic calculus of institutional change outlined in Chapter 2. Efforts to coordinate water governance arrangements across jurisdictional boundaries are constrained when transaction costs of coordination outweigh the benefits, presenting a challenge to reduce the transaction costs through institutional innovations, and to accept that fragmentation persists for other reasons, including legitimacy and representation. Basin closure implies that the 'costs of non-cooperation' are rising in a context of upstream–downstream interdependencies (Molle et al., 2010). Therefore, the benefits of collaboration and coordination must be sufficiently high and fairly distributed to justify the costs borne by a 'winning coalition' of states and stakeholders needed to pass the reform. In the absence of such a coalition among the states in the Murray–Darling, the federal government invoked new authority and committed substantial financial and administrative resources to overcome the resistance from selected states. The increasing concentration of authority in the federal government under the 2007 Commonwealth Water Act can be viewed as an effort to reduce the high costs of consensus-based interstate decision-making (McKay, 2012; Connell, 2011b, 2011a). However, the backlash triggered by the federalization of basin planning and allocation has revealed trade-offs. Although transaction costs of consensus are high, the transaction costs of implementing the decisions might be made lower due to the increased ownership engendered through reaching consensus.

The theory of institutional collective action (Feiock, 2013) recognizes that there are multiple institutional mechanisms for coordination or integration ranging between the extreme positions of central control versus chaotic fragmentation governed exclusively by informal networks. Marshall et al. (2013) argue that the creation of a federal Basin Authority is an example of a formal, or 'overt', mechanism to address institutional collective action dilemmas. Overt measures include administrative processes for watershed planning, formal deliberation and purchaser–provider relationships. They argue that such overt mechanisms are '*no longer* able to overtly coordinate the suite of interdependent enterprises relevant to the success of water management efforts in the region' (p. 214). In contrast, 'covert' measures involve informal collaboration and competitive rivalry and are meant to capture the emergent properties of complex, polycentric and multilayered governance arrangements. This distinction between the overt and covert aligns with Feiock's discussion of the typology of integration mechanisms based on their level of formal authority and their scope.

Is there a way to make sense of this mixture of overt and covert approaches to integrated water market and basin governance reforms? To navigate and map the complex institutional landscapes governing water markets and integrated river basin management, the next section of the chapter conceptualizes river basins as examples of local public economies, introducing key concepts about public goods and transaction costs, following Schlager and Blomquist (2008) and Oakerson and Parks (2011), and further elaborating the typology of integration mechanisms in relation to the collaboration risks and transaction costs of the institutional collective action dilemmas associated with efforts to scale up integrated water markets in the Murray–Darling.

A LOCAL PUBLIC ECONOMIES VIEW: WHEN THE 'LOCAL' BECOMES A LARGE RIVER

In their seminal paper on polycentric governance of metropolitan areas, Vincent Ostrom et al. (1961) noted that cities do not need to be consolidated or centralized into a single unit; instead, a city with multiple, independent centers of authority can be effective as long as there are coordination mechanisms in place to ensure governing bodies operate coherently as a system. Such is the fine line between order and chaos in polycentric governance arrangements. Fragmentation without coordination can lead to chaos. But fragmentation confers many potential advantages associated with decentralization and distribution of powers, such as access to local knowledge and values, the ability to experiment without failure spreading system-wide, and potential for policy diffusion based on successful local innovations. The key is to couple fragmentation with coordination around important values and goals (Pahl-Wostl et al., 2012). Coordination comes in many forms, from working groups and purchaser–provider contracts to regional authorities, and is not synonymous with centralization that could override the benefits of polycentric governance arrangements. These fine but crucial distinctions hold several insights for the Murray–Darling where constitutional reservation of authority for the states has fragmented authority. In this context, path dependency has contributed to the accretion and accumulation of different integration mechanisms over time where new forms are developed without fully eliminating their precursors, leading to unpredicted impacts. This is exemplified by The Living Murray first step, which relied on consensus-based decision-making before being replaced by the Commonwealth Environmental Water Office mantra of 'cooperative but independent' (ANOA 2011).

Theories of local public economies are closely related to the concept of polycentric governance and were developed initially to understand the complex organization of metropolitan governance (Ostrom et al., 1961; Oakerson, 1999). Despite the once-dominant prescription to govern cities based on centralized delivery of public services, empirical studies illustrated that alternative arrangements – which may appear fragmented and chaotic on the surface – could perform coherently and as well as or better than their more centralized counterparts. Prescriptions for comprehensive, centrally planned resource management within bioregional boundaries is analogous to the calls for centralized municipal governance in the 1950s and 1960s. This has made the theory of local public economies relevant to analyze and explain patterns of decision-making about public goods and their delivery in large-scale common pool resource systems such as river basins and ecoregions (Schlager and Blomquist, 2008; Oakerson and Parks, 2011; Marshall, 2009).

There are three conceptual pillars of the theory of local public economies (Oakerson and Parks, 2011). First, the provision of public goods is distinct from their production. Public goods involve two components: (1) provision: a decision to provide a public good; and if so, a choice about (2) production: a means of producing the good in question. Provision refers to decisions about the quantity and quality of goods and services to provide publicly (versus privately), regulation, revenue generation and 'how to arrange for the production of goods and services' (Oakerson, 1999). Provision decisions are driven in turn by voter and stakeholder preferences, including the agencies responsible for provision who may seek rents that interfere with the principle of fiscal equivalency: the notion that beneficiaries pay for goods and services. Production, on the other hand, refers to the 'technical transformation' of resources into outputs, which depends on time- and place-specific information, which implies the need for different economies of scale based on the goods and services in question (Oakerson, 1999). For example, the decisions and funding to return the Murray–Darling Basin (MDB) to sustainable levels of extraction are (potentially) separate from the institutional arrangements used to deliver water for the environment. Provision and production are linked, however, via contracting or other types of coordination mechanisms.

Second, local public economies deliver multiple public goods for a diverse constituency and, hence, can become quite complex. The formation of specialized governance units does not necessarily lead to chaotic fragmentation, however; instead, there is potential for complex organization and coherent patterns of behavior and coordination even in competitive situations. This relates closely to the distinction between overt and covert mechanisms for integrated water resource management in the Murray–Darling, as elaborated by Marshall et al. (2013), and noted above; there is potential for coherent interstate water management based on 'competitive rivalry and informal collaborations among enterprises within a polycentric public industry, as well as self-organized efforts by enterprises to resolve their conflicts using available political or legal instruments' (Marshall et al., 2013). Such is the premise guiding recent proposals to grant water allocations to non-profit conservation groups to deliver water for local environmental needs, as a response to demands for localism to involve local communities and non-governmental actors in environmental watering (Robinson et al., 2014).

Finally, there is an emphasis on the empowerment of communities of interest and citizens in provision decisions on the one hand, and competition among production units to stimulate public entrepreneurship on the other. The tension between cooperation and competitive rivalry in local public economies seems to mirror precisely the uneasy balance between capping and trading, respectively. However, this tension also characterizes other instances of horizontal coordination, that is, when states cooperate within intergovernmental agreements but must compete for federal resources.

Although upfront investments in transition costs of coordination are needed to reduce transaction costs over the longer term, transaction costs are ultimately a limiting factor in local public economies, curbing both possible extremes of unification within one consolidated governance unit (Oakerson, 1999) and unfettered proliferation of public goods provision and production (Oakerson and Parks, 2011: 149–50): 'The multiplicity of provision units is constrained theoretically by transaction costs . . . When permitted to do so, citizens can be expected to create a new unit when they believe the benefits anticipated exceed the costs of creation and operation; provision units would not simply proliferate'.

Basin governance arrangements have the potential to generate different mixtures of public goods and services based on the politics, stakeholder preferences and coordination mechanisms involved. In this context, the roles and responsibilities of users, states and the Commonwealth government are continually being renegotiated, and integration has proven elusive, particularly within a comprehensive and unified framework. Governance arrangements have evolved in relation to a growing bundle of water-related public goods; an inventory of these goods provides the basis for analyzing the relationship between institutional design, transaction costs and adaptive efficiency (Schlager and Blomquist, 2008) in the ongoing reform process in the Murray–Darling. This builds on the discussion in Chapter 2 about the status of water as a complex economic good with its multiple interconnected public and private values (Hagedorn, 2008), including:

- water supply reliability;
- water resource development and energy production;
- water quality;
- risk management for drought and flooding;
- rural economic development and structural adjustment;
- environmental water recovery.

INTEGRATION MECHANISMS: TYPOLOGIES AND TRENDS

Each of the public goods defined above is associated with a community of interest whose boundaries are based on the preferences of a 'group of people who share some contiguous part of the local geography'; the boundaries of the 'local' vary based on the nature of the problem and public good in question (Oakerson, 1999). Often these communities overlap. Different communities of interest may compete with one another either to use scarce water resources for different public goods or to use water for private goods (irrigation) instead of a public good (environmental flows) (cf. Schlager and Blomquist, 2008; Lankford and Hepworth, 2010). The challenge of integration is at root one of reconciling the diverse communities of interest that have a stake in the different public goods associated with water. Basin closure has required increasing trade-offs between the multiple public and private goods delivered by rivers, reinvigorating debates about the institutional design to deliver these multiple outcomes and make trade-offs across them.

What institutional form should integration take? In which circumstances is coordination at the basin scale necessary or desirable, and how should roles and responsibilities be distributed and coordinated? The proliferation of social, ecological and economic values associated with water and rivers underpins institutional collective action dilemmas for multiple jurisdictions and stakeholders with at least two orders of dilemmas: a first-order dilemma connected to prioritizing which public goods to deliver; and then a second-order dilemma connected to delivering that good, akin to the provision and production distinction established by Oakerson.[5]

Institutional responses to these dilemmas have been described as 'integration mechanisms' (Feiock, 2013), where integration can be viewed in terms of the capacity of polycentric governance arrangements to operate coherently as a system (Ostrom et al., 1961). This becomes the operational criterion for adaptive efficiency of polycentric governance arrangements at the intersection of water trading and river basin governance reforms. Integration mechanisms span a diverse range from informal working groups to regional authorities with formally defined powers and functions. Feiock (2013) has established a typology of integration mechanisms characterized by two major attributes: authority and scope. Authority can be either informal or formal. It varies from informal networks to increasingly formal arrangements: contracts, delegated authority and imposed authority. All but the latter (imposed authority) rely on voluntary cooperation among jurisdictions and communities of interest. Scope can be a single functional area or comprehensive. It ranges from a single issue and bilateral agreements to comprehensive coverage of multiple functions.

Mechanisms also vary in terms of their coordination costs, which tend to increase as their scope expands and authority becomes more formalized. Feiock (2013: 399) notes that 'transaction costs are viewed as a

Table 5.3 Integration mechanisms for institutional collective action dilemmas: the Murray–Darling reform experience

Scope	Authority		
	Embeddedness	Contracts	Delegated authority
Encompassing, complex	*Multiplex self-organizing system*	*Councils of governments* Murray Darling Basin Commission/ Ministerial Council (1987/92); Council of Australian Governments Reform (1994)	*Regional authorities* State governments; Murray–Darling Basin Authority (2007)
Intermediate, multilateral	*Working groups* Independent Audit Group (cap); COAG water resource policy	*Partnerships* Murray–Darling Basin Agreement (1987/92); National Water Initiative (2004)	*Multipurpose districts* Catchment management authorities
Single-issue, bilateral	*Informal networks*	*Service contracts* Water trading; Operations (delivery, salinity control, etc)	*Single-purpose districts* Irrigation organizations; Environmental water holders

Note: Imposed authority is an important integration mechanism in (re)centralizing governance arrangements, such as the Australian Murray–Darling; however, it is not an example of voluntary collective action and therefore not included with the matrix of integration mechanisms noted above. As noted by Feiock (2013: 401), integration mechanisms associated with imposed authority 'are not necessarily a product of collaboration'.

Source: Based on Feiock (2013).

primary barrier to mitigating ICA dilemmas'. As noted above, the evolution of intergovernmental integration mechanisms in the Murray–Darling Basin can be viewed in terms of the costs of non-cooperation (that is, benefits of cooperation) and the transaction costs of coordination (Table 5.3). Considering the typology of transaction costs elaborated in Chapter 2 (following Basurto and Ostrom, 2009; Garrick et al., 2013), the adoption of integration mechanisms is limited to situations in which the benefits of coordination outweigh the associated costs of developing, adopting and implementing a rule change, in both the short and the long term. This calculus may be conducted through the biased lens of specific political interests or bureaucratic politics which include or exclude key categories of costs and benefits to justify the status quo or a shift to a desired administrative arrangement. Central to Feiock's argument is the notion that more formal and encompassing arrangements will be reserved for only those situations with the highest 'collaboration risk', where states and other jurisdictions threaten to defect, divide resources or benefits unfairly, or fail to coordinate on critical matters. These are precisely the challenges associated with cap-and-trade in the Murray–Darling Basin because water trade, diversion limits and environmental water recovery involve a range of distributional concerns about who wins and who loses.

Table 5.3 adapts Feiock's typology of integration mechanisms using the two main dimensions: scope and authority. The typology can be used to capture the evolving public economy of river basin governance in the Murray–Darling and the proliferation and interaction of integration mechanisms. Consider authority, which can be informal, contractual, delegated and imposed. On one end of the spectrum, authority for integration mechanisms is informal and embedded. Informal networks, working groups and multiplex self-organizing systems are three types in this category, varying in their scope from narrow to broad. Irrigators may coordinate certain operational and maintenance functions among neighbors through informal networks without recourse to formal rules, although this has become increasingly formalized as competition has intensified and markets have matured.

As authority becomes more formal, contracts are used to structure voluntary negotiation or intergovernmental agreements. Examples include service contracts, partnerships and councils of governments. Partnerships and intergovernmental agreements have become prevalent for the Murray–Darling dating from the River Murray Commission and, particularly, the Murray–Darling Agreement in 1987 and 1992. The 2004 intergovernmental agreements for the National Water Initiative and The Living Murray are other prominent examples. The creation of a council of governments may or may not involve new statutory authority but their scope is typically broader and more comprehensive. The Council of Australian Governments, for example, developed a wide-ranging water reform package as a component of a sweeping National Competition Policy.

Delegated authority is the most formal and ranges in scope from single or multi-purpose districts to regional authorities with formal powers. The establishment of semi-autonomous irrigation organizations is an example of delegated authority but with the scope limited to irrigation matters. The Murray–Darling Basin Authority (MDBA) was established by the 2007 Water Act as a federal authority organized along regional, basin-wide boundaries. The Basin Authority is an example of a regional authority with relatively broad powers. Even though the Water Act carefully circumscribed the MDBA's scope, it assigned important functions to establish diversion limits and identify environmental water requirements, which entailed several tasks.[6] This remit depended on recourse to different constitutional provisions because the water allocation authority is reserved for the states under section 100 of the Constitution (Fisher, 2011; Skinner and Langford, 2013).

Considered in terms of Feiock's typology, the evolution of integration mechanisms in the Murray–Darling has produced two outcomes. First, there are multiple layers of integration mechanisms – both formal and informal, general and specific in scope – that overlap, adding redundancy but also potential confusion due to interactions with preexisting arrangements. Although some of the layers (for example, working groups) have fixed lifespans, many continue indefinitely and accumulate. Second, there is a trend toward increasingly formal and encompassing water rights and basin governance arrangements, marked most recently by the establishment of new federal authority, in some cases crowding out informal networks and intergovernmental partnerships. This federalization of basin governance in the interest of efficiency (narrowly defined) may have come at the expense of legitimacy and resilience (which are inherent to adaptive efficiency).

REVISITING THE SCALING-UP CHALLENGE

Recasting the challenge of scaling up as a challenge of institutional collective action opens new analytical terrain to trace the evolution of polycentric governance and integration mechanisms in the Murray–Darling (Marshall et al., 2013) for: market-oriented water rights reform, basin governance to develop and adjust the cap; and to acquire and deliver water for the environment, as well as manage the interactions and trade-offs between water trade and environmental water recovery. The Australian National Water Commission (National Water Commission, 2011a, 2011b) documents three periods of water market reforms: emergence, expansion and a transition to sustainability. Polycentric governance arrangements for water rights reform have evolved across these three periods to enable trade and ensure its compatibility with sustainable and adaptive water planning and allocation institutions.

The sections below briefly consider the public goods at stake, and how they have changed across these periods, tracing the implications for provision, production, transaction costs and integration mechanisms. The first period (pre-1992) promoted water security for irrigation systems as the primary public good, and relied on state and federal provision of infrastructure, subsidies and preliminary changes to licensing systems as limits were reached in the 1970s and 1980s. Cross-jurisdictional integration was limited to the activities of the River Murray Commission.

The second period (1992–2007) relied on intergovernmental agreements and the integration mechanisms of the Murray–Darling Basin Commission to maintain historic irrigation production and halt the decline in environmental services. The Murray–Darling Basin Commission offered the intergovernmental apparatus for provision by state and Commonwealth governments with production coordinated by irrigation infrastructure organizations, catchment management authorities and nascent intergovernmental environmental watering programs under The Living Murray. Ad hoc working groups were nested within the overarching framework and coordination provided by the Commission.

The third period – and current arrangements – rely on new federal authority to return the basin to sustainable levels of extraction while limiting impacts on irrigation production: the Murray–Darling Basin Authority and Commonwealth Environmental Water Office (ultimately grounded in the Constitution's section 51 'external affairs' powers). Resistance to these new federal roles in turn triggered a localism initiative involving regional organizations and non-governmental contributions to environmental water delivery. Table 5.4 summarizes these periods.

Polycentric arrangements for water rights reform have evolved through sequential efforts to create benefits and share risks at the basin level. To provide context for this trajectory, these three phases can be considered alongside the water trading patterns since the early 1980s when water and land rights began to be separated (see Figure 5.1). These trading patterns are juxtaposed with federal expenditures associated with environmental water recovery during the transition to sustainable diversion limits. Lee and Ancev (2009) document approximately AU$25 billion across the COAG reforms, 2008 Water for the Future Program, and a series of other federal expenditures. An additional AU$1.7 billion was committed in 2012 and earmarked for irrigation efficiency improvements.[7] The increase in federal expenditures reflects the growing challenges of achieving water reform goals that require coordination across disparate communities of interest surrounding irrigation, states and the environment. These investments in institutional transitions at the federal level (a form of transaction costs associated with decision-making and monitoring), coupled with property rights reform in states and irrigation districts, enabled trading. These investments represented upfront transaction costs with an aim to

Table 5.4 Public good/CPRs, transaction costs and polycentric integration mechanisms

Phase	Public good/CPRs	Dominant transaction costs	Polycentric integration mechanisms
Emergence	• Irrigation • 'Drought-proofing'	*Institutional transitions:* Defining tradable property rights; establishing trading rules; reporting and registries *Static transaction costs:* Administrative fees (state and irrigation); brokerage	Regional authority (state)
Expansion	• Irrigation • Drought risk sharing • Environmental recovery	*Institutional transitions:* Establishing the cap; harmonizing water rights; revising trading rules; defining environmental requirements *Static transaction costs:* Reporting, administrative fees; irrigation organization termination fees	Working groups; partnership agreements, councils of governments, river basin commission (interstate)
Transition to sustainability	• Irrigation • Environmental recovery • Structural adjustment	*Institutional Transitions:* Basin Plan to adjust the sustainable diversion limits; trading rules to coordinate water trade and environmental recovery *Static transaction costs:* Reporting, administrative fees; irrigation organization termination fees	Imposed authority; additional partnership agreements; additional working groups; Commonwealth government

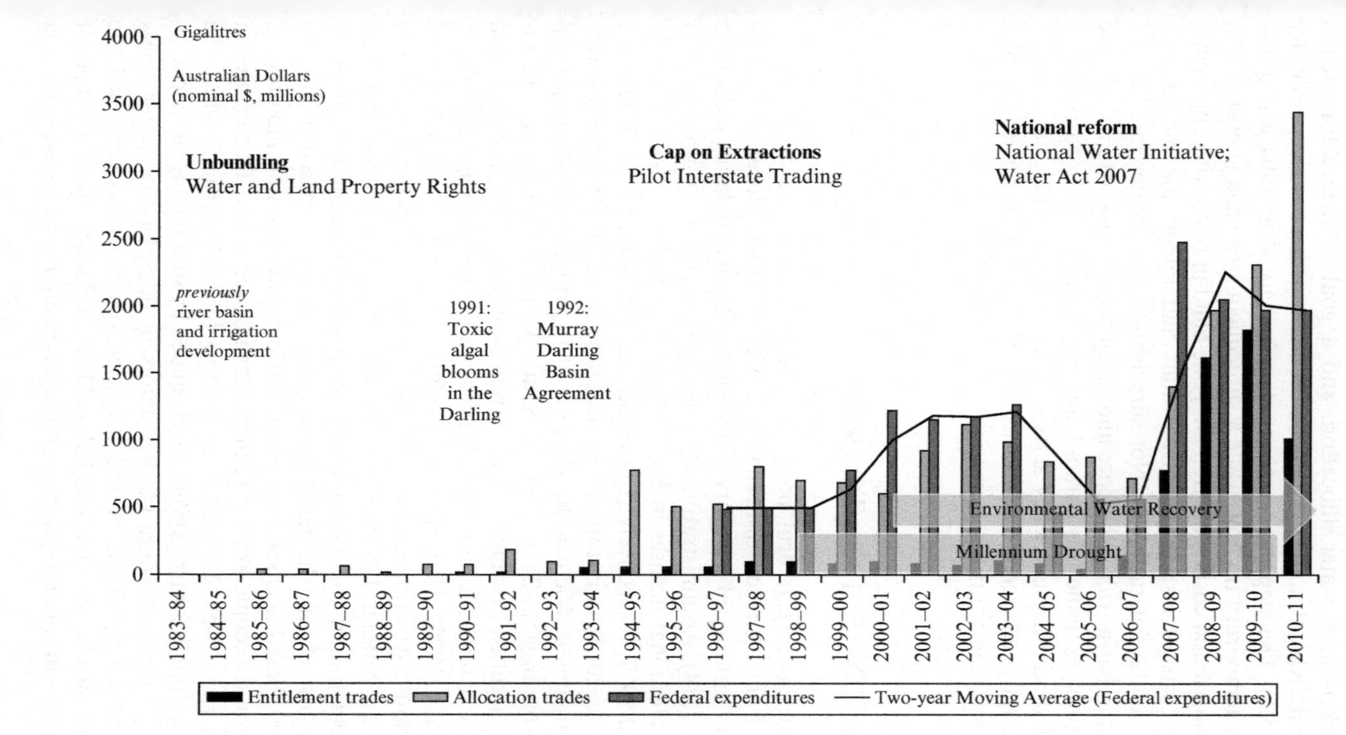

Source: Garrick et al. (2013).

Figure 5.1 Water trade and federal expenditures on the transition to sustainable diversion limits

strengthen the institutional framework for coordinated basin planning and market-based water allocation, and a goal to reduce transaction costs for individuals and consumptive users competing for access to water (Figure 5.1). The step increase in trading activity after 1994 was dominated by temporary trading, while a subsequent step change occurred in 2007. Transaction costs reduction has gone hand in hand with apparently growing costs of institutional transitions and integration mechanisms (at least at the federal level). Not surprisingly, the greater interest and attention in basin-wide coherence and integration comes with a cost. The next sections examine this paradox and its implications for institutional design and adaptive efficiency across the three periods in greater depth, where adaptive efficiency is concerned with the sequentially more complex investments in institutional transitions to maintain or decrease transaction costs over time (cf. Chapter 2).

EMERGENCE: 1970s TO 1992

The contemporary reform process originated with state governments, flowing from their constitutional authority in water allocation matters and over a century of statutory water rights reforms predating federation (Tisdell, 2014). Administrative decisions by state water resource agencies curbed additional licenses in South Australia and New South Wales in 1969 and 1977, respectively (Bjornlund and O'Callaghan, 2003; Grafton and Horne, 2014); moratoria were introduced for some (physically) unregulated streams of Victoria by the late 1970s and early 1980s (National Water Commission, 2011a, 2011b). South Australia went beyond moratoria: it modified existing licenses to reflect historic use patterns instead of eligible acreage, leading to reductions in license volumes (Bjornlund, 2003; IAG, 1996). Permanent and temporary trading was introduced first in South Australia in 1982 and then temporary trading followed in New South Wales in 1983 (IAG, 1996). Victoria approved temporary and permanent trading in 1987 and 1991, respectively. During this phase of water rights reform, the states started to convert area-based licenses to volumetric-based licenses to enable easier measurement and monitoring, which would become key for reducing transaction costs. This change in property rights was also aimed at closing a loophole through which the land area eligible for irrigation was expanded by subdividing property.

State governments were the key actors in this phase of water rights reform; however, irrigators and irrigation organizations remained the dominant producers and wielded political influence over a range of decisions about water development. Irrigation interests resisted moratoria on licenses and sought to protect subsidies for infrastructure and operations. Land and water rights were bundled until the mid-1980s, which preserved the connection between water licenses and irrigation. Trading activity was thin and typically included transfers of licenses connected with land sales. Several years elapsed between moratoria on water licenses and active trading. Transition costs could therefore be measured in the time period (years) between moratoria and trading activity. For example, in South Australia – the first state to approve trading in 1982 – there was an average of 38 permanent water transfers per year between 1987 and 1993 (Bjornlund and McKay, 1998). By the time of the COAG reforms of 1994, permanent trading in Victoria totalled 15 gigalitres (GL) or 0.7 per cent of entitlements, and temporary trading only began to expand in 1994–95 to 200 GL or 8 per cent of water usage, the vast majority of trade involving rights that had never been used previously.

Integration mechanisms for interjurisdictional coordination remained very limited. Trading concentrated within irrigation districts. State governments performed a coordination role within their territory but experienced substantial technical challenges in harmonizing property rights, monitoring and enforcement criteria across irrigation districts. These coordination challenges plagued efforts to establish integrated water markets within states, let alone across them. The River Murray Commission – the body established in 1917 to implement the 1914–15 Murray Waters Agreement – had narrow formal authority tied to construction and operation of the dams and weirs, and implementation of the interstate water sharing agreements. Although the Commission fell short of its full potential for intergovernmental cooperation (Clark, 1971), it offered a platform for ad hoc cooperation by the states on matters outside the narrow remit of the Murray Waters Agreement, such as data collection, through what Feiock might term a 'multiplex self organizing system' governed by informal norms across a potentially wide scope of activities outside of the Commission's formal remit.

By the late 1980s, water rights reform was uneven; trading was cumbersome and transaction costs intensive. Complex transactions involved the purchase of land to secure appurtenant water entitlements, followed by the reassignment of water to different parcels within the property (Bjornlund and O'Callaghan, 2003). Benefits of water trade were emerging slowly in an era of limits, yet irrigation communities resisted market-based water rights reforms due to the costs and equity concerns. The implementation lag was also a consequence of relatively abundant water availability and loopholes (including access to unused water licenses). During the period from 1988 to 1993 immediately preceding the Coalition of Australian Governments reform process (described below), diversions totalled 10680

GL and only 63 per cent of total water entitlements were used (IAG, 1996). Further expansion of water trading would require a coherent and consistent capping system along with compatible water planning arrangements, separation of land and water rights, and the reduction of barriers to trade between irrigation communities and across jurisdictions. These reforms would depend in turn on a changing political economy and calculus of the benefits and costs of institutional reform, particularly recognition of basin closure due to hot weather, low flows and salinity (1991–92 Darling algal bloom) and drought (1994–95 and 1997–2009), coupled with declining water supply alternatives (as previously unused licenses known as sleepers and dozers became activated).

EXPANSION: 1992 TO 2007

Intensified scarcity and the emergence of new environmental values prompted water reforms to achieve more productive and sustainable water use. Transition and transaction costs of water rights reform included changes to water licensing systems (described below) and the creation of a vast network of trading rules coordinated first within districts, then states and, most recently, within the Southern Murray and entire Murray–Darling system, estimated to include more than 30 000 trading rules (Waterfind, 2013). These transitions depended on renewed intergovernmental cooperation to achieve river-level integration for the first time since negotiations surrounding the River Murray Waters Agreement in the early 1900s.

Market-based water rights reforms (trade) and basin-wide governance arrangements (cap) were viewed not only as compatible but also as interdependent: 'The Cap per se, is only a means to an end. It is not the end in itself. The overall objectives can be achieved only by identifying environmental water requirements and flow regimes and by establishing a supporting management and institutional framework, including trading of water' (IAG, 1996: 41).

This period of expansion and broadening of water trade coincided with and was propelled by new integration mechanisms with additional scope and authority: the intergovernmental Murray–Darling Basin Agreement. A series of working groups also proved instrumental during this period. In Feiock's institutional collective action framework, working groups rely on informal authority matched to narrowly defined problems with a relatively limited potential for opportunism. Working groups proved well equipped to tackle key elements of the reform process but insufficient to harmonize water rights and trading rules across irrigation districts and states. The Murray–Darling Basin Commission and Ministerial Council provided the oversight to identify such gaps and inconsistencies, even in water allocation domains for which it lacked formal authority to act without interstate consensus. This reflects a mutualism between working groups and intergovernmental agreements between the states.

At least five key working groups shepherded the reform process through this water market broadening phase, illustrating the opportunities and limits for informal mechanisms for integration:

- water resource policy group of the Coalition of Australian Governments;
- Independent Audit Group commissioned by the Murray–Darling Basin Ministerial Council (MDBMC);
- working group on a property right in water;
- water market reforms group (trading rules); and
- interstate pilot trading group.

In 1994, the COAG kicked off this process with a water reform package based on the recommendations of the COAG working group on water resources policy led by Sir Eric Neal. The essential focus of the COAG reform was to apply competition policy to the water sector by ensuring water charges are based on cost recovery and financing is generated for rural infrastructure and its maintenance. Much has been written about the focus on minimizing transaction costs to remove impediments to reallocation. The working group also diagnosed 'a lack of clear definition concerning the role and responsibilities of a number of institutions involved in the industry' (COAG, 1994). New modes of state and intergovernmental cooperation were needed to reform the irrigation industry, property rights systems and water measurement. In so doing, 'where they have not already done so, governments would develop administrative arrangements and decision-making processes to ensure an integrated approach to natural resource management' (COAG, 1994: 5).

In 1995, a second working group established a Framework on the Implementation of a Property Right in Water (ARMCANZ, 1995) which identified the elements of 'efficient' property rights. Water and water rights are: (1) 'in demand'[8]; (2) well specified in the long-term sense; (3) exclusive; (4) enforceable and enforced; and (5) transferable and divisible. These principles accord with the theoretical prescriptions promoting efficient property rights that are exclusive, enforced and transferable with low transaction costs (Cheung, 1969), also known as '3D' property rights: defined, defensible and divestible (Bougherara et al., 2008). Translating these principles for private tradable rights into practice entailed new regulations and information systems, as well as administrative capacity, at the state, interstate and Commonwealth levels. The National Competition Council was tasked in 1995 with tracking progress and performance of the reforms. The working group on property rights reform struggled to establish consistency across states, which limited early trading within districts and states, particularly in the absence of a coherent and consistent cap.

In response, the Murray–Darling Basin Ministerial Council commissioned a third working group in July 1996, the Independent Audit Group (IAG), to review state proposals for establishing an interim cap on water diversions. The IAG was guided by six principles to ensure equity and consistency across the states: (1) no additional deterioration of ecological flow regimes; (2) a precautionary approach to water allocation that limits ecological harm; (3) allocative efficiency to ensure water is allocated to its highest-valued (economic) use; (4) recognition of statutory property rights; (5) transparent and auditable water management processes; and (6) that administrative efficiency based on 'a system of administration be implemented which is easily understood and which minimizes time and costs' (IAG, 1996: ix). This led to a hierarchy of property rights that prioritized rights based on statutory commitments, a history of use, and/or firm promises to future access. The upstream states of New South Wales and Victoria agreed to limit diversions to the levels under 1993–94 levels of infrastructure development; South Australia was permitted to develop its high security entitlements, which would allow for a small increase over 1993–94 diversions. Queensland's cap would be based on an audited water planning process. However, this hierarchy was not imposed on a blank slate: each state operated with its own licensing system. Despite efforts of the IAG to harmonize the cap across states, New South Wales sought a unique standard, and Queensland and the Australian Capital Territory lagged.

As a result of differences in the cap and licensing systems, as well as policies in Victoria and New South Wales that restricted water exports, interstate trade also lagged. This posed particular problems for downstream South Australia. An interstate water trade pilot started in 1998, once again on the basis of a working group formed under the umbrella of the Murray–Darling Basin Commission (MDBC) and MDBMC. The pilot trading involved 55 trades of 9.5 GL during the initial two-year period from 1998 to 2000. An audit of the program revealed that brokers preferred to address administrative problems before extending the pilot. One stark indication of the transaction costs were the 32 days required for permitting forms to travel by post (Young et al., 2000). The audit also underscored the need for streamlined administration through common forms and more direct communication between regulators in different states. A Pricewaterhouse Coopers (2006) study commissioned by the

Department of the Prime Minister concluded that 'robustness, liquidity and efficiency of market arrangements were less-than-ideal' and 'inter-state trade has developed at a slower pace reflecting complexities associated with trading across jurisdictional borders, such as differences in entitlement specification and trading rules'. States and irrigation organizations jealously guarded against water leaving their jurisdictions. In one oft-cited example, Victoria imposed a rule to limit permanent trade out of irrigation regions to less than 4 per cent of irrigation system water use per year to ease the transitions within regional economies, known as 'trading rule 25' under the 1989 Victoria Water Act. This rule has only been phased out in 2014, more than 15 years since the pilot program commenced.

GROWING PAINS AND TRANSACTION COSTS REDUCTION: 2004–07

Two intergovernmental agreements in 2004 established new integration mechanisms to remedy dysfunctions as markets broadened: the National Water Initiative and The Living Murray agreement (the latter is covered in the next section). The National Water Initiative sought to harmonize water rights systems and statutory water plans across the river valleys in the basin. In so doing, it sought to streamline statutory water planning, continue water rights unbundling across a range of attributes (water delivery, water use, storage) and streamline the growing number of trading rules. It also assigned water shortage risks between irrigators and states.

The National Water Initiative also established the National Water Commission (NWC) to assess impacts of water market reforms and coordinate statutory water planning processes at the state level. The NWC began its operations in 2005 and has evaluated the efficiency of property rights based on more detailed criteria. A fifth working group – the water market reform group – set out a system of trading rules. By 2006, a brokerage, Waterfind, estimated that there were more than 30 000 rules in place to facilitate trading in the MDB (Allen Consulting Group, 2006: 9). The Australian Competition and Consumer Commission, formed to promote competitive industry across a range of sectors in the Australian economy, has become integral to this process and provides more formal oversight and direction for rules. During the Basin Plan, it developed a new set of trading rules to preserve the competitiveness of the market in the context of new sustainable diversion limits.

The criteria established by the 2004 National Water Initiative (NWI) were operationalized by the National Water Commission in its *Strengthening*

Australia's Water Markets report (NWC, 2011a) to identify seven major categories, which align with key actors and the integration mechanisms used to address ICA dilemmas of a maturing market (Table 5.5). These criteria demonstrate the complementarity and overlap in integration mechanisms to couple narrowly defined working groups (informal) or service contracts (formal) within the formal and more encompassing arrangements provided by the National Water Initiative, Murray–Darling Basin Commission and the Ministerial Council.

By 2006, water trade and interstate began to pick up (see Figure 5.1). The National Water Initiative and preceding reforms were effective in scaling up water trade, while drought created new pressure and opened potential gains from trade. Gains from trade have been estimated using general equilibrium models to tease out the net effects of water trade on national and regional gross domestic product (GDP), totalling AU$1.36 billion from 2006 to 2012 in the Southern Murray–Darling (National Water Commission, 2010). Observations have confirmed these benefits. Kirby et al. (2014) reviewed the role water trade played in the context of the Millennium Drought: by 2006, when the critical winter inflows were 10 per cent of historic average and storages were depleted by the accumulated effects of drought, water allocations were reduced by almost 70 per cent in the 2008–09 water year compared to a baseline of 2000–01. Gross value in irrigated agriculture decreased by only 20 per cent in commodity price adjusted terms. Water trade was cited as critical in enabling productivity increases (in economic output per unit of water) and allocative efficiency (by allowing water to move from relatively low-value annual crops such as cotton and rice to higher-valued perennial plants for horticulture and grapes). These benefits accrue across the basin states of this region. The upstream states (particularly Victoria) captured the vast majority of this stream of benefits estimated by modelling, while observed trading patterns during drought saw a net export of 500 GL from upstream jurisdictions to downstream South Australia (National Water Commission, 2011b). This illustrates the distributional aspects of water trade and the associated interjurisdictional coordination challenges.

During this period of growing pains, clause 58 of the National Water Initiative acknowledged the need 'to minimize transaction cost on water trades, including through provision of good information flows in the market, and compatible entitlement, registry, regulatory and other arrangements across jurisdictions'. This came on the heels of increasing efforts to document transaction costs across private and public actors in the market. In his study of water trading of the late 1990s in the South Australian Murray, Challen modelled transaction costs and estimated the out-of-pocket expenditure of AU$70 per megalitre for administrative

Table 5.5 Enabling mechanisms for water trade and transaction costs reduction

Mechanism	Elements	Key actors	Integration mechanisms
Rights definition and titling	Clearly defined water entitlements; unbundled water rights (access, use and delivery); statutory water plans; statutory public water registers; clear risk assignment framework	Irrigation organizations and states	*Partnership* Intergovernmental Agreement for National Water Initiative
Market and trading rules	Market rules to meet specific objectives; charge rules (termination fees, transaction fees); defined processes for rule changes	Australian Competition and Consumer Commission	*Service contracts* >30 000 trading rules; coordinated by regional authority (MDBA)
Governance	Separation of regulatory, policy, commercial and operational functions; appeals process for regulatory decisions; transparent process for water allocation determinations	States	*Council of Governments* Australian Competition and Consumer Council
Market information	Publication of real time storage and inflow data; accurate and timely reporting of trade volume and price data; coordinated release of water allocation determinations	Bureau of Meteorology and MDBA	*Regional authority*
Compliance and enforcement	Statutory water accounting; water metering; cost-effective verification and audit systems; appropriate penalties for non-compliance	States and MDBA	*Regional authority*
Market administration and trade processing	Electronic lodgment of applications; transparent approvals processes	States	*Intergovernmental agreement* (NWI standards)
Market intermediaries	Competitive market for intermediate services	Private	Private

Source: Based on NWC (2011a).

(up to AU$50) and brokerage (AU$20) fees with additional implicit costs, with total transaction costs corresponding to about 3 to 29 per cent of an average trade. The Allen Consulting Group (2006) estimated a subset of both the transition and transaction costs. Transaction costs for market participants (so-called 'private' costs) included an application fee (up to AU$500), registration of trade (up to AU$130), taxes (stamp duty and income taxes) and other governmental costs (for example, assessments), as well as any brokerage fees on both the buyer and seller side. Transaction costs vary across states, temporary transactions typically incur lower costs than permanent, and South Australia's fees were among the highest.

Transaction costs were no longer prohibitive or as cumbersome as the pre-COAG period, declining to an estimated 2.7 per cent of the value of temporary water traded in Victoria, to 21 per cent in South Australia, demonstrating an apparent decrease since the work by Challen. Processing times were as fast as a day for temporary trade in Victoria and typically no longer than a week in the other states. Although permanent trades still took far longer, and up to six months in New South Wales, permanent trade had begun to increase by 2007 (see Figure 5.1). In July 2009, the Council of Australian Governments and Natural Resource Management Ministerial Council adopted processing standards of approving 90 per cent of allocation trades within five (intrastate) and ten (interstate) days, respectively, in all states but South Australia where standards were slightly longer. The standards simultaneously demonstrate the success of the past investments by working groups and interstate partnerships and councils in reducing transaction costs (processing times), and also the continued need for a portfolio of integration mechanisms to reinvest in planning and water rights reforms needed for adaptive efficiency.

State governments were important underwriters of the transition and transaction costs, such as the AU$7 million cost over four years for Victoria to unbundle its water rights, and the AU$8.5 million annual expenditures in New South Wales used in part to convert its water licenses (Allen Consulting Group, 2006). The states also bear the lion's share of contributions from governments to the operations of the Murray–Darling Basin Commission and its successor, totalling approximately AU$100 million per year. This illustrates how the integration mechanisms needed for broadening water markets are costly. This unlocked gains from trade for irrigators and opened new opportunities for environmental restoration.

The 2007 Water Act and 2012 Basin Plan re-established the goal to 'minimize transaction costs of water trades, including through good information flows in the market and compatible entitlement, registry, regulatory and other arrangements across jurisdictions' (Australian Government, 2007). This goal exists alongside the commitment to enable trading within and across states, enable new water market products, recognize and protect the environment, and protect third party impacts. The new trading rules adopted in 2014 are the latest effort to continue this trajectory of transaction costs reduction to enable gains from trade across a growing range of water market products (for example, temporary and permanent trade, but also lease arrangements, trade in storage rights known as carryover, and even futures trading). These rules decrease transaction costs by providing information and reporting (for example, requiring trading prices to be declared), removing restrictions on trade (for example, volumetric limits based on point of origin or destination), and so on. States are required to publish their trading rules in a central repository to aid transparency and enable interstate trade.

TRANSITION TO SUSTAINABILITY: 2007 TO PRESENT

The water governance arrangements established by the 2007 Water Act and 2012 Basin Plan can be traced to the implementation lags of consensus-based interstate integration mechanisms: the National Water Initiative and The Living Murray First Step. The NWI enabled water trade to scale up, as evidenced by the increase in 2007–08, yet the commitment to sustainability and environmental recovery lagged. The Living Murray (TLM) program, adopted in 2004 in parallel with the NWI, was an intergovernmental agreement with a AU$500 million state cost-share to acquire water for the environment for six internationally important ecological 'icon sites'. Staff members from the state and Commonwealth agencies involved in initial planning, acquisition and delivery activities had a generally positive perception of TLM (Garrick et al., 2012). However, the deliberative process for interstate consensus required time and tested the patience of many observers and politicians in TLM's initial years of operation.

The Water Act 2007 restructured authority and established two new integration mechanisms: the Murray–Darling Basin Authority (MDBA) (Basin Plan) and the Commonwealth Environmental Water Office, two examples of 'imposed authority' in Feiock's scheme, marking a departure from voluntary institutional collective action among the states and Commonwealth. Figure 5.2 documents the contributions from governments to the annual budgets for the Murray–Darling Basin Commission (until 2007) and the Murray–Darling Basin Authority since 2008. After a steady increase in contributions from governments, a shift in state–federal funding for the Murray–Darling Basin Commission foreshadowed this

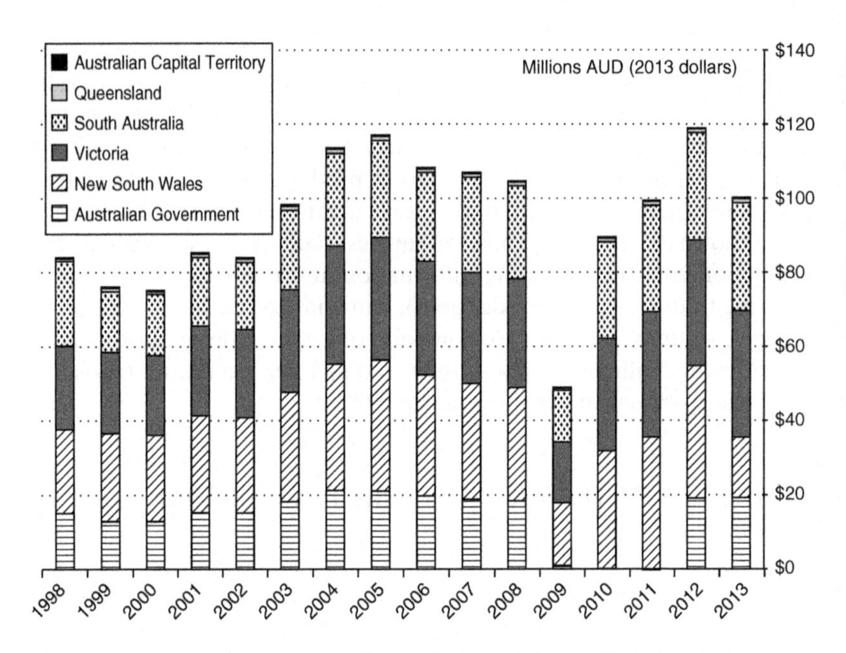

Source: Annual Reports, Murray–Darling Basin Commission (until 2007) and Murray–Darling Basin Authority (since 2008).

Figure 5.2 Murray–Darling Basin Commission and Authority: contributions from governments to annual budgets

restructuring from the consensus-based MDBC to the federal MDBA with unilateral authority. In addition to the annual contributions from government noted in Figure 5.2, an AU$500 million grant from the Commonwealth government to the MDBC was made on 19 May 2006 to 'facilitate timely implementation of pre-existing decisions of both the Council and Commission' and 'accelerate water recovery measures' (Australian Government, 2008). This infusion epitomized the wider trend to concentrate authority and power at the Commonwealth level to overcome the inertia and high transaction costs of interstate coordination under the consensus model.

In 2007, former Liberal Prime Minister John Howard proposed a reform package in a politically charged period of drought response. The AU$10 billion Howard Plan for Water Security was the initial blueprint for the Commonwealth Water Act and drafted by Environment Minister Malcolm Turnbull. The Act granted the Murray–Darling Basin Authority with authority to determine environmental water requirements and thereby adopt sustainable diversion limits. This represented a fundamental shift, from consensus-based decision-making and intergovernmental agreements among the states within the Murray–Darling Basin Ministerial Council, to unilateral action. On the face of it, this contravened Section 100 of the Constitution and would normally depend on the referral of state powers to the federal government. However, Victoria withheld its support to seek additional federal money in exchange for its lost control over water allocation (see Skinner and Langford, 2013; Kildea and Williams, 2011). Instead, the Commonwealth derived constitutional authority through the Section 51 external affairs clause on the basis of the Act's environmental objectives to fulfil international wetlands treaties, including the Ramsar Convention. This set the conditions ripe for constitutional challenge, and increased the collaboration risk and potential for opportunism by states throughout the implementation of the Act (Webster and Williams, 2012).

The Basin Plan was the 2007 Water Act's integration mechanism to define sustainable diversion limits based on environmental water requirements. The 2008 Water for the Futures Program (an Amendment to the Commonwealth Water Act) earmarked AU$3.1 billion for buybacks of water entitlements in addition to almost AU$6 billion for irrigation efficiency projects, and water recovery began while the Plan was being developed (Productivity Commission, 2010). After a contentious political debate, the Plan was adopted by Parliament with an overall recovery target of 2750 GL for the environment, a permanent 20 per cent reduction from historic diversion levels. Approximately 450 GL in additional water recovery was committed by the Commonwealth in 2012 to meet South Australia's demands for increased water security through irrigation efficiency investments that would appease Victoria and New South Wales. As of late 2013, 1337 GL had been recovered through buybacks with a long-term average yield (reliability) of 1137 GL. An additional 710 GL have been recovered through irrigation efficiency measures at considerably higher cost than direct buybacks (Crase et al., 2013; Marshall, 2013). The latter may be considered the political price of the recovery process, with the full costs yet to be accounted for due to new investments in such measures only strengthening the lock-in to infrastructure approaches inherited from past water policy decisions. It further signalled a sea change in environmental water recovery, from the use of water plans to recover water for the environment, to the acquisition and delivery of entitlements for the environment (National Water Commission, 2014). As a consequence the distinction or boundaries between water trade and basin planning and coordination have blurred. In this context, environmental water recovery has led to a complex governance challenge involving new federal agencies, as well as state governments, catchment organizations, irrigation associations and non-profit organizations (Garrick et al., 2012). Fragmented governance has been recognized as a central obstacle in environmental water acquisition and delivery, and hence raises many institutional collective action challenges (Productivity Commission, 2010; Cull et al., 2011). Institutional fragmentation and the management of trade-offs across jurisdictions motivated the vertical integration of environmental watering decisions within the Commonwealth Environmental Water Office (CEWO). However, political resistance encountered during the implementation of the Act coupled with the complexities of environmental watering have elevated the importance of a local role to coordinate irrigation and environmental uses. The CEWO's 2009 decision-making framework for environmental watering also reflects this wider trend toward localism; stakeholders emphasized the need to 'clarify institutional arrangements and involve local communities in the delivery of environmental water' (Department of Environment, 2009). A 2011 parliamentary inquiry also scrutinized community consultation efforts by the MDBA (Windsor Committee, 2011).

Integration mechanisms evolved from consensus-based partnership agreements to what the Australian National Audit Office (2011) describes as 'cooperative-but-independent' Commonwealth authorities vested in the CEWO and MDBA. In practice, scaling up water trade and environmental recovery has entailed polycentric governance challenges and a new phase in the evolution of integration mechanisms. The Commonwealth Environmental Water Office was created by the Water Act as a federal entity to hold and manage environmental water reserves, yet it depends increasingly on states, service contracts and working groups across all elements of decision-making and implementation. Therefore the CEWO's polycentric arrangements demonstrate a gap between the integration mechanisms on paper – an imposed authority – and those in practice – a series of working groups, contracts, partnerships, and so on. Skinner and Langford (2013) identify 17 programs for environmental water recovery, with the integration mechanisms evolving from state-level agencies, catchment management entities or non-profits to more formalized authority and expansive scope in the form of the MDBA and CEWO, before shifting to a mixture of formal and informal mechanisms with both comprehensive and issue-specific scope, for example the agreement between the CEWO and Nature Foundation South Australia to manage up to 10 GL per year for locally established priorities (Robinson et al., 2014).

The governance arrangements for environmental water recover and associated water rights reform have produced a diverse institutional landscape marked by dynamic tension between subsidiarity and complementarity (Garrick et al., 2012; Marshall, 2008): a combination of integration mechanisms linking locally coordinated capacity (horizontal) with cross-scale linkages across tiers of governance (vertical).

CONCLUDING REMARKS

This chapter defined river basin closure as a dilemma for institutional collective action and argues that a diverse portfolio of integration mechanisms are needed to 'scale up' water allocation reforms across three elements: water trading, capping and environmental water recovery. Scaling up does not mean centralization, however; instead, scaling up is irreducibly polycentric in practice, if not also in form. Framing these vertical and horizontal coordination dilemmas in terms of institutional collective action follows in the recent tradition advanced by Marshall et al. (2013) in the context of the Murray–Darling, Schlager and Blomquist (2008) in the Western US, and Feiock (2008) for situations of intergovernmental public goods provision more broadly. In so doing, new analytical and theoretical perspectives are now available to understand why fragmentation persists in the Murray–Darling, as well as the diverse portfolios of integration mechanisms that reconcile opposing forces in polycentric governance arrangements: subsidiarity and complementarity, overt and covert mechanisms, formal and informal arrangements, and issue-specific and multi-dimensional coordination.

Water reform in the Murray–Darling has produced an uncomfortable, yet largely fruitful, marriage of water markets and integrated river basin governance that is held together by a bundle of integration mechanisms of varying authority and scope. River basin closure has expanded the boundary of water's 'local' public economy from the irrigation community to the basin as a whole, and even beyond the basin to the dependent cities, agricultural markets and international conservation initiatives. Contrary to the premise of free market environmentalism, water trading is far from a self-maintaining system that unlocks decentralized responses to price signals as water scarcity intensifies. However, a central authority for integrated river basin governance has not been a panacea either. Despite the swing of the pendulum from consensus-based intergovernmental arrangements to new forms of imposed authority or overt coordination mechanisms with the 2007 Water Act and 2012 Basin Plan, there remains a weak track record of compliance and a low likelihood that this trend will change, due to high transaction costs of interjurisdictional coordination for complex, multi-dimensional public goods at multiple scales from the community level to basin scale. Instead, there is emerging evidence that unilateral imposed authority and broadly encompassing mechanisms threaten to crowd out important informal and targeted working groups and consensus-based intergovernmental partnerships, which work slowly but build legitimacy and momentum toward adaptive efficiency. Meinzin-Dick underscores the importance of getting beyond panaceas in water institutions, which is apt in this context, noting the strength of an 'institutional tripod' that combines states, markets and community-based management, rather than a single option in isolation. The experience of the Murray–Darling illustrates that a 'Goldilocks' compromise is needed, lying between the extreme positions of decentralization, on the one hand, and increasing concentration of authority in central government, on the other. Such arrangements couple intergovernmental partnership agreements and regional authorities with local capacity and informal working groups to achieve adaptive efficiency based on the complementarities. This has been driven home by the unilateral retreat from key planks of the National Water Initiative, such as the decision by the new Liberal government led by Tony Abbott to eliminate funding for the National Water Commission and associated reporting and coordination mechanisms.

NOTES

1. Although coordination should not be confused with centralization.
2. Jurisdictions may pursue their goals to the detriment of common interests.
3. Typically defined and assessed in neoclassical terms as the maximization of economic welfare in a static sense, in contrast with the concept and criterion of adaptive efficiency discussed in Chapters 2–4 in this volume, and revisited later in this chapter.
4. As noted below in greater detail, Feiock (2013: 397) describes 'institutional collective action dilemmas' as those that 'arise directly from the division or partitioning of authority in which decisions by one government in one or more specific functional area impact other governments and other governmental functions'.
5. Thank you to Graham Marshall for pointing out this important distinction.
6. The MDBA (2012) 2011–12 Annual Report identifies its functions as including to construct and operate dams and weirs, develop the Basin Plan, advise the Commonwealth Water Minister on the accreditation of state water resource plans, develop an information platform for water trading, share water between the states, 'manage all aspects of Basin water resources', and carry out measurement and monitoring, information gathering, community engagement and education.
7. The New South Wales government reduced its contributions by more than half in an effort to undermine the Basin Plan and the recovery process (Australian Broadcast Corporation, 2014).
8. Quantity available is 'scarce relative to the amount demanded'.

REFERENCES

AGRICULTURE AND RESOURCE MANAGEMENT COUNCIL OF AUSTRALIA AND NEW ZEALAND STANDING COMMITTEE ON AGRICULTURE AND RESOURCE MANAGEMENT (ARMCANZ). 1995. Water Allocations and entitlements: a national framework for the implementation of property rights in water. Policy Position Paper for discussion with stakeholders in developing and implementing systems of water allocations and entitlements. In: TASKFORCE ON COAG WATER REFORM (ed.), *Occasional Paper Number 1*. Canberra: Council of Australian Governments.

ALLEN CONSULTING GROUP. 2006. Transaction costs of water markets and environmental policy instruments. Report to the Productivity Commission.

ANDERSON, T.L. and LEAL, D.R. 2001. *Free Market Environmentalism*, New York, Palgrave Macmillan.

AUSTRALIAN NATIONAL AUDIT OFFICE (ANAO) (2011). Restoring the Balance in the Murray–Darling Basin. ANAO Audit Report No. 27 2010–11. Available at: http://www.anao.gov.au (accessed March 2012).

AUSTRALIA PARLIAMENT HOUSE OF REPRESENTATIVES STANDING COMMITTEE ON THE ENVIRONMENT AND HERITAGE. 2000. *Co-ordinating Catchment Management: Report of the Inquiry into Catchment Management*. Canberra: Australian Government.

AUSTRALIAN GOVERNMENT. 2007. Australia (Commonwealth) Water Act 2007, Act no. 137 as amended. Available at: http://www.comlaw.gov.au/Details/C2014C00715.

BARK, R.H., PEETERS, L.J.M., LESTER, R.E., POLLINO, C.A., CROSSMAN, N.D. and KANDULU, J.M. 2013. Understanding the sources of uncertainty to reduce the risks of undesirable outcomes in large-scale freshwater ecosystem restoration projects: an example from the Murray–Darling Basin, Australia. *Environmental Science and Policy*, 33, 97–108.

BAUER, C.J. 2004. *Siren Song: Chilean Water Law as a Model for International Reform*. Washington, DC: Resources for the Future.

BELL, S. and QUIGGIN, J. 2008. The limits of markets: the politics of water management in rural Australia. *Environmental Politics*, 17, 712–29.

BERKES, F. 2002. Cross-scale institutional linkages: perspectives from the bottom up. In: STONICH, S., STERN, P.C., DOLSAK, N., DIETZ, T., OSTROM, E. and WEBER, E.U. (eds), *The Drama of the Commons*. Winnipeg: National Academies Press.

BJORNLUND, H. 2003. Efficient water market mechanisms to cope with water scarcity. *International Journal of Water Resources Development*, 19, 553–67.

BJORNLUND, H. and MCKAY, J. 1998. Factors affecting water prices in a rural water market: a South Australian experience. *Water Resources Research*, 34, 1563–70.

BJORNLUND, H. and O'CALLAGHAN, B. 2003. *Property Implications of the Separation of Land and Water Rights*. Brisbane: Pacific Rim Real Estate Society (PRRES).

BLOMQUIST, W.A., DINAR, A., HAISMAN, B. and BHAT, A. 2005. Institutional and policy analysis of river basin management: the Murray–Darling River Basin, Australia. World Bank Policy Research Working Paper.

BOUGHERARA, D., GROLLEAU, G. and MZOUGHI, N. 2008. The 'make or buy' decision in private environmental transactions. *European Journal of Law and Economics*, 27, 79–99.

CHALLEN, R. 2000. *Institutions, Transaction Costs, and Environmental Policy: Institutional Reform for Water Resources*. Cheltenham, UK and Northampton, MA, USA: Edward Elgar Publishing.

CHEUNG, S.N. 1969. Transaction costs, risk aversion, and the choice of contractual arrangements. *Journal of Law and Economics*, 12, 23–42.

CLARK, S. D. 1971. River Murray Question: Part II-Federation, Agreement and Future Alternatives, The. *Melb. UL Rev.*, 8, 215.

CONNELL, D. 2007. *Water Politics in the Murray–Darling Basin*. Annandale, NSW: Federation Press.

CONNELL, D. 2011a. The role of the Commonwealth environmental water holder. In: CONNELL, D. and GRAFTON, R.Q. (eds), *Basin Futures: Water Reform in the Murray–Darling Basin*. Canberra: ANU ePress.

CONNELL, D. 2011b. Water reform and the federal system in the Murray–Darling Basin. *Water Resources Management*, 25, 3993–4003.

CONNELL, D., DOVERS, S. and GRAFTON, R.Q. 2005. A critical analysis of the National Water Initiative. *Australasian Journal of Natural Resources Law and Policy*, 10, 81–107.

CONNELL, D. and GRAFTON, R.Q. 2011. Water reform in the Murray–Darling Basin. *Water Resources Research*, 47, W00G03-9.

CRASE, L., O'KEEFE, S. and DOLLERY, B. 2011. Some observations about the reactionary rhetoric circumscribing the guide to the Murray–Darling Basin Plan. *Economic Papers*, 30, 195–207.

CRASE, L., O'KEEFE, S. and DOLLERY, B. 2013. Talk is cheap, or is it? The cost of consulting about uncertain reallocation of water in the Murray–Darling Basin, Australia. *Ecological Economics*, 88, 206–13.

CULL, S., PLOSKO, S., BYRNES, A. and CASS, B. 2011. Restoring the balance in the Murray–Darling Basin. In: AUSTRALIAN NATIONAL AUDIT OFFICE (ed.), *Audit Report*. Canberra: Australian National Audit Office.

DEPARTMENT OF ENVIRONMENT, WATER, HERITAGE and ARTS. 2009. A Framework for Determining Commonwealth Environmental Watering Actions. DEWHA, Canberra.

EBC, RMCG, MARSDEN JACOB ASSOCIATES, ECONSEARCH, MCLEOD, G., CUMMINS, T., ROTH, G. and CORNISH, D. 2011. *Community Impacts of the Guide to the Proposed Murray–Darling Basin Plan. Volume 2: Methodology. Report to the Murray–Darling Basin Authority*. Canberra: Murray–Darling Basin Authority.

FEIOCK, R.C. 2013. The institutional collective action framework. *Policy Studies Journal*, 41, 397–425.

FISHER, D. 2011. A sustainable Murray–Darling Basin: the legal challenges. In: CONNELL, D. and GRAFTON, R. (eds), *Water Reform in the Murray–Darling Basin*. Canberra: ANU E-Press.

GARRICK, D., BARK, R., CONNOR, J. and BANERJEE, O. 2012. Environmental water governance in federal rivers: opportunities and limits for subsidiarity in Australia's Murray–Darling River. *Water Policy*, 14, 915–36.

GARRIDO, A. 1998. An economic analysis of water markets within the Spanish agricultural sector: can they provide substantial benefits? In: EASTER, K.W., ROSEGRANT, M. and DINAR, A. (eds), *Markets for Water: Potential and Performance*. New York: Kluwer.

GRAFTON, R.Q. and HORNE, J. 2014. Water markets in the Murray–Darling Basin. *Agricultural Water Management*, 145, 61–71.

GRAFTON, R.Q., LIBECAP, G., MCGLENNON, S., LANDRY, C. and O'BRIEN, B. 2011. An integrated assessment of water markets: a cross-country comparison. *Review of Environmental Economics and Policy*, 5, 219–39.

HAGEDORN, K. 2008. Particular requirements for institutional analysis in nature-related sectors. *European Review of Agricultural Economics*, 35, 357–84.

HARRIS, E. 2007. Historical regulation of Victoria's water sector: a case of government failure? *Australian Journal of Agricultural and Resource Economics*, 51, 343–52.

HARRIS, E. 2011. The impact of institutional path dependence on water market efficiency in Victoria, Australia. *Water Resources Management*, 25, 4069–80.

INDEPENDENT AUDIT GROUP (IAG). 1996. *Setting the Cap: Report of the Independent Audit Group*. Canberra: Murray–Darling Basin Commission.

KILDEA, P. and WILLIAMS, G. 2011. The Constitution and the management of water in Australia's rivers. *Sydney Law Review*, 32, 595–616.

KIRBY, M., BARK, R., CONNOR, J., QURESHI, M. E. and KEYWORTH, S. 2014. Sustainable irrigation: How did irrigated agriculture in Australia's Murray–Darling Basin adapt in the Millennium Drought? *Agricultural Water Management*, 145, 154–62.

LANKFORD, B. and HEPWORTH, N. 2010. The cathedral and the bazaar: monocentric and polycentric river basin management. *Water Alternatives*, 3, 82–101.

LEE, L.Y.-T. and ANCEV, T. 2009. Two decades of Murray–Darling water management: a river of funding, a trickle of achievement. *Agenda: A Journal of Policy Analysis and Reform*, 16, 5.

LICHATOWICH, J. 2001. *Salmon Without Rivers: A History of the Pacific Salmon Crisis*. Washington, DC: Island Press.

LOCH, A., WHEELER, S., BEECHAM, S., EDWARDS, J., BJORNLUND, H. and SHANAHAN, M. 2013. The role of water markets in climate change adaptation. Adelaide: National Climate Change Adaptation Research Facility and University of South Australia.

MARSHALL, G. 2008. Nesting, subsidiarity, and community-based environmental governance beyond the local scale. *International Journal of the Commons*, 2, 75–97.

MARSHALL, G.R. 2009. Polycentricity, reciprocity, and farmer adoption of conservation practices under community-based governance. *Ecological Economics*, 68, 1507–20.

MARSHALL, G.R. 2013. Transaction costs, collective action and adaptation in managing complex social–ecological systems. *Ecological Economics*, 88, 185–94.

MARSHALL, G., CONNELL, D. and TAYLOR, B. 2013. Australia's Murray–Darling Basin: a century of polycentric experiments in cross-border integration of water resources management. *International Journal of Water Governance*, 1, 197–218.

MCKAY, J. 2012. The theory and practice of Australian institutional reforms to incorporate water markets in integrated water resources management. *Water Trading and Global Water Scarcity: International Experiences*, 259.

MOLLE, F., WESTER, P. and HIRSCH, P. 2010. River basin closure: processes, implications and responses. *Agricultural Water Management*, 97, 569–77.

MURRAY–DARLING BASIN AUTHORITY (MDBA). 2010. Keeping salt out of the Murray. Canberra: MDBA.
MURRAY DARLING BASIN COMMISSION. 2007. Annual Report 2006–7. Canberra: Murray Darling Basin Commission.
NATIONAL WATER COMMISSION (NWC). 2010. The impacts of water trading in the southern Murray–Darling Basin: an economic, social and environmental assessment. Canberra: NWC.
NATIONAL WATER COMMISSION (NWC). 2011a. *Strengthening Australia's Water Markets*. Canberra: ACT.
NATIONAL WATER COMMISSION (NWC). 2011b. *Water Markets in Australia: A Short History*. Canberra: NWC.
NATIONAL WATER COMMISSION (NWC). 2014. *Australian Environmental Water Management: 2014 Review*. Canberra: ACT.
NEUMAN, J.C., SQUIER, A. and ACHTERMAN, G. 2006. Sometimes a great notion: Oregon's instream flow experiments. *Environmental Law*, 36, 1125–55.
OAKERSON, R.J. 1999. *Governing Local Public Economies: Creating the Civic Metropolis*. Oakland, CA: ICS Press.
OAKERSON, R.J. and PARKS, R.B. 2011. The study of local public economies: multi-organizational, multi-level institutional analysis and development. *Policy Studies Journal*, 39, 147–67.
OSTROM, V., TIEBOUT, C.M. and WARREN, R. 1961. The organization of government in metropolitan areas: a theoretical inquiry. *American Political Science Review*, 55, 831–42.
PAHL-WOSTL, C. and KNIEPER, C. 2014. The capacity of water governance to deal with the climate change adaptation challenge: Using fuzzy set Qualitative Comparative Analysis to distinguish between polycentric, fragmented and centralized regimes. *Global Environmental Change*, 29, 139–54.
PAHL-WOSTL, C., LEBEL, L., KNIEPER, C. and NIKITINA, E. 2012. From applying panaceas to mastering complexity: toward adaptive water governance in river basins. *Environmental Science and Policy*, 23, 24–34.
POWELL, J.M. 1989. *Watering the Garden State: Water, Land, and Community in Victoria, 1834–1988*. Sydney: Allen & Unwin.
PRODUCTIVITY COMMISSION. 2010. Market mechanisms for recovering water in the Murray–Darling Basin. *Final Report*. Canberra.
PUJOL, J., RAGGI, M. and VIAGGI, D. 2006. The potential impact of markets for irrigation water in Italy and Spain: a comparison of two study areas. *Australian Journal of Agricultural and Resource Economics*, 50, 361–80.
QURESHI, M.E., SHI, T., QURESHI, S.E. and PROCTOR, W. 2009. Removing barriers to facilitate efficient water markets in the Murray–Darling Basin of Australia. *Agricultural Water Management*, 96, 1641–51.
ROBINSON, C., BARK, R., GARRICK, D., POLLINO, C. (2014). Sustaining local values through river basin governance: community-based initiatives in Australia's Murray–Darling Basin. Journal of Environmental Planning and Management. DOI: 10.1080/09640568.2014.976699.
SCHLAGER, E. and BLOMQUIST, W.A. 2008. *Embracing Watershed Politics*. Boulder, CO: University Press of Colorado.
SIEMANN, D. and MARTIN, S. 2007. Managing many waters: an assessment of capacities for implementing water and fish improvements in the Walla Walla Basin. Pullman, WA: William D. Ruckelshaus Center.

SKINNER, D. and LANGFORD, J. 2013. Legislating for sustainable basin management: the story of Australia's Water Act (2007). *Water Policy*, 15, 871–94.
TISDELL, J. 2014. The evolution of water legislation in Australia. In: EASTER, W.K. (ed.), *Water Markets for the 21st Century*. Dordrecht: Springer.
WATERFIND. 2013. CEO Report. Adelaide: Waterfind.
WEBSTER, A. and WILLIAMS, J. 2012a. Can the High Court save the Murray River? *Environmental and Planning Law Journal*, 29, 281–96.
WEBSTER, A.L. and WILLIAMS, J.M. 2012b. Can the High Court save the Murray River? *Environmental and Planning Law Journal*, 29, 281–96.
WINDSOR COMMITTEE. 2011. Of drought and flooding rains: inquiry into the impact of the guide to the Murray–Darling Basin Plan. Canberra: House of Representatives, Standing Committee on Regional Australia.
YOUNG, M.D. and MCCOLL, J. 2008. A future-proofed basin: a new water management regime for the Murray–Darling Basin. Adelaide: University of Adelaide.
YOUNG, M., MCDONALD, D.H., STRINGER, R. and BJORNLUND, H. 2000. *Interstate Water Trading: A Two Year Review*. Canberra: CSIRO Publishing.
YOUNG, O.R. 2002. *The Institutional Dimensions of Environmental Change: Fit, Interplay, and Scale*. Cambridge, MA: MIT Press.
ZETLAND, D. 2011. *The End of Abundance: Economic Solutions to Water Scarcity*. Amsterdam: Aguanomics Press.

6. Systemic risks, polycentric responses: performance, principles and practices

> Even though federal and state policy fosters the export of agricultural commodities, Western water law generally inhibits trade in the water used to grow the commodities. States should open up the market by eliminating or streamlining legal barriers that effectively block transfers of water. . .The Western water crisis is basically an imbalance between supply and demand. Opening water resources to trade has the potential to reduce the imbalance by rewarding water conservation, ensuring that water goes toward the highest-value and most-efficient uses, and providing the financial tools to mitigate fluctuations in water availability.
>
> (Robert Glennon and Gary Libecap, 2014, Wall Street Journal)

INTRODUCTION

In March 2015, droughts in Brazil and California captured the news headlines as this book went to press. Declining reservoirs in Sao Paulo and California exposed systemic risks and interdependencies in river basins and aquifers under pressure from chronic stresses and extreme climate events. Systemic risks refer to 'breakdowns in an entire system' rather than its constituent parts; such risks are defined by the prevalence of modest tipping points that produce cascading failures, trigger a chain of losses and reduce resilience to future shocks (Kaufman and Scott, 2003). River basin closure poses systemic risks because even minor or moderate disturbances can produce unanticipated, or underestimated, consequences.

Interest in water markets has been renewed with predictable vigor in a context of mounting demand and more volatile supplies. Water trading remains an attractive alternative to costly supply-side solutions, maladaptive water allocation or unsustainable depletion of groundwater reserves. Therefore, 40 years after the National Water Commission heralded water transfers as part of the 'future of water policy', it is reasonable to wonder if this future has finally arrived? This is particularly germane because climate change is projected to bring even more frequent and severe droughts, including 'mega-droughts' longer than any observed in the historical record (Cook et al., 2015). Demand is projected to increase to meet the needs of an urbanizing and developing world with rising standards of living.

Notwithstanding these pressures, for every investment in water allocation reforms, such as California's 2014 groundwater law, there is a major investment in engineering solutions (for example, components of the $7.5 billion water bond in California passed the same year). This suggests that expensive supply-side options are still appealing, or at minimum deemed feasible, as illustrated by the water importation options considered for the 2012 Colorado River Supply and Demand Study (US Bureau of Reclamation, 2012). The options for importing water to the Front Range of Colorado, the Green River Headwaters and Southern California have capital costs of at least $3 billion per scheme and annual unit costs from $700 per acre foot to $3400 per acre foot. These expenses are high but they are more widely distributed than the third-party costs associated with water transactions. The political economy of water allocation reform suggests that it is not only the magnitude of the costs, but their distribution, that affect the feasibility of water markets, and hence the transaction costs associated with addressing the equity concerns involved.

If the future of water policy is now, and water trading is to (continue to) play a part, it comes with a new and pragmatic view of the role of transaction costs in institutional reform. For example, Peter Culp et al. (2014) produced a report on *Shopping for Water* to advance water markets in the Western US. They note that the high transaction costs generated by 'existing procedural and regulatory restrictions' in the Western US are 'not necessarily undesirable or inappropriate, even if they result in fewer trades than might otherwise occur' (p. 13). This echoes the hypothesis from Colby (1990: 1189) a generation earlier that 'the transaction costs generated by state regulation of water transfers out of agriculture are lower than would be socially optimal' because they fail to adequately account for the pervasive externalities and interdependencies associated with transfers.

It is one thing to recognize that allocating water involves substantial transaction costs via any means, including water trading. It is another to confront the implications for theory and practice. In this volume, I have argued that the path toward integrated water markets involves substantial, sustained and multilevel investments in governance capacity, including the need for water rights reform to be paired with coordination institutions at multiple scales. This explains the arrival of a new era of planning and river basin governance, illustrated by 22 basin studies in the Western US since the passage of the 2009 Secure Water Act. These system-wide planning efforts assess long-range supply and demand trends in a changing climate. They facilitate critical interstate and multistakeholder information-gathering, deliberation and coordination to scale up reforms, even if incrementally; water transactions are only one tool to address the supply–demand imbalances. Markets and systemic, multilayered planning go hand in hand to reconcile the private and public interests in water allocation. The high transaction costs incurred in the process are – at least partially – the symptoms of efforts to secure legitimacy amidst contested values.

PERFORMANCE: TRANSACTION COSTS AND ADAPTIVE EFFICIENCY

This book has exposed the institutional underpinnings of water markets by examining the evolution, performance and interaction of water rights reform and river basin institutions in three rivers under pressure. A core lesson of this book is that transaction costs can never be eliminated, or even rendered trivial. Minimizing some types of transaction costs (particularly the transition costs of deliberative decision-making and coordination institutions) can unravel progress over the long term. In a world of positive and substantial transaction costs, adaptive efficiency is an overarching goal, which is by definition an elusive and moving target. Adaptive efficiency refers to trajectories of economic performance in the face of pervasive uncertainties, systemic risks and shocks, feedbacks and tradeoffs across multiple scales. Paradoxically, adaptive efficiency requires a concern for criteria other than efficiency, including equity, adaptability and robustness.

Adaptive efficiency can only be assessed over the long term, implying both a historical approach and a consideration of the future. In the context of path dependency, adaptive efficiency involves 'sequentially more complex' institutional innovations to lower, but not eliminate, the costs of transacting (Carey and Sunding, 2001: 291). The accumulated evidence suggests that the quest for adaptive and sustainable water allocation reform remains elusive after decades of such sequences of investment. Even when the costs of transacting have decreased, or if the benefits of transacting justify the expense, the costs of the institutional innovations themselves remain high and possibly on the rise, requiring new analytics to understand the interplay of transition and transaction costs in the evolution of water allocation reform across multiple scales.

What can be said about the trajectories of institutional development and performance of water allocation reforms in the three rivers? All three rivers have achieved incremental progress in the face of increasing pressures from growing demand, development and drought. Strictly in terms of efficiency, the Murray–Darling has established the most advanced water market in the world with approximately 30 per cent of water allocations (annual allotments) being traded per year in the Southern Murray–Darling, with an annual boost to regional GDP of up to $370 million according to general equilibrium modelling (Grafton and Horne, 2014; National Water Commission, 2011). The static transaction costs of water trading have declined markedly, with approvals requiring less than a week and direct costs limited primarily to administrative fees. Trading activity is more limited in the Colorado and Columbia Rivers, constituting a much smaller fraction of total water use, and far less than the trading potential identified by many observers. Transaction costs are substantial and have not been systematically tracked and measured. In the state of Colorado, for example, one of the initial transactions implemented by the Colorado Water Trust involved over $70 000 in donated legal and hydrological expertise (Malloch, 2005), while transaction costs represented up to 70 per cent of total costs in the environmental water transactions of the Upper Columbia Basin from 2003 to 2007 (Garrick et al., 2013). Reductions have occurred in places such as the Deschutes, Lemhi and Big Thompson-Colorado, but they remain the exception.

Accounting for sustainability, equity and adaptability, the performance trends are decidedly more mixed both today and in terms of the trajectories of change integral to adaptive efficiency. In terms of sustainability criteria, the Murray–Darling has gone the farthest – in volumetric and percentage terms – toward restoring ecosystem health. Moreover, it has adopted and adapted sophisticated diversion limits based on ecological and social criteria. On the other hand, the Colorado has yet to acknowledge the arrival of hard limits; several experts have now argued that the recognition and regulation of limits are a critical priority, based directly on exchanges between academics and practitioners in the Colorado and Murray–Darling. In the Columbia, limits have been adopted via multiple institutional mechanisms, including court decrees, administrative rules, and so on. Progress has been patchy, and implementation is even more limited.

On concerns of equity and legitimacy, and by extension the adaptability of water allocation institutions, all three rivers attempted to reform water allocation institutions to confront the omissions of the past – environmental needs, cities, groundwater and indigenous water claims – to paraphrase Kenney (2009). Addressing these issues has raised additional equity concerns for the historic users and interests bearing the concentrated costs associated with water allocation reforms: irrigators and some subnational jurisdictions (states) either slow to develop or with less secure water apportionments. The resulting efforts to establish regulatory safeguards and multijurisdictional plans and river basin institutions have met with limited success and fragile compromises; even in the best instances they are saddled by high transaction costs (of coordination) and low compliance.

The Murray–Darling had an auspicious start to its basin governance reforms with the Murray–Darling Basin Initiative, yet efforts to accelerate and centralize this process have threatened lasting progress. A polycentric governance mixture of formal and informal institutions has proven critical to enhance adaptive efficiency in the aftermath of the 2012 Basin Plan. The Colorado has fostered a period of deal-making and provisional interstate compromise despite a history of legal conflict; conversely, the Columbia has established strong connections between water rights reform and local watershed governance in four US states, although progress is uneven and limited by a lack of binding coordination institutions. These experiences in the Colorado and Columbia reflect different solutions to the equity concerns associated with water allocation reform, in part reflecting the unique history, starting points and institutional actors involved. The combination of efficiency, sustainability and equity gives an impression of the adaptability and adaptive efficiency of the water allocation institutions in the three rivers. The Murray–Darling initially outpaced the Western US rivers, but the Colorado has achieved steady progress as pressures intensify, evidence of adaptive efficiency as an emerging property of institutional diversity. The Columbia still strives to scale up and has experimented with novel public–private partnerships in the Columbia Basin Water Transactions Program in the absence of strong coordination institutions.

The outcomes of water allocation reform in the three rivers therefore need to be viewed against the counterfactuals (that is, what were the alternatives?) to 'restore a sense of the options that confronted people at the time to show the grit and friction that was evident in every one of these episodes' (Blackbourne, 2006: 12). In the Colorado, predictions of Compact calls and legal conflicts have failed to materialize (yet); instead, drought has fostered an era of unprecedented cooperation between states, with complex new instruments being devised to experiment and learn from. In the Columbia, after decades of court cases, legal reform and rule-making, a steady trickle of water has been acquired for the environment, which in some cases has led toward more comprehensive sub-basin planning. Institutional reforms to scale up and coordinate basin-wide outcomes within a Columbia Basin Water Transactions Program are under way, to harness and spread pockets of innovation. In the Murray–Darling, it is more than ten years since the 'First Step' of The Living Murray in 2004 committed 500 GL for the environment. Despite the concerns of retrenchment and retreat from this pathway to water reform, the basin has achieved a step shift in both the goals of the reform – to achieve sustainable diversion limits – and implementation progress, with more than 1500 GL in environmental holdings and a maturing water market with sophisticated rules to coordinate consumptive and environmental water allocation.

THEORETICAL IMPLICATIONS: PRINCIPLES FOR INSTITUTIONAL DESIGN

It is no longer enough to acknowledge that we live in a world of positive transaction costs, nor is it enough to argue simply for flexibility in an uncertain and contested institutional landscape for water allocation. Instead, I have examined the evolution of institutions for collective action to address two linked dilemmas: the allocation of common pool water resources and the provision of the multiple public goods that underpin them at several scales. Chapter 1 described river basin closure as a large-scale collective-action dilemma with corresponding challenges of complex property systems, high coordination costs and cross-scale governance. There are some opportunities to scale up the lessons from small-scale commons, but there are limits. This points to the need for theoretical and analytical approaches to examine multilevel collective action in a dynamic context.

Chapter 2 introduced concepts, analytical perspectives and emerging evidence about water allocation in a transaction costs world. Although multiple influential theoretical traditions have assessed transaction costs and property rights in natural resource allocation, none is sufficient on its own to understand the evolution and performance of water allocation policy. Blending perspectives from Coase, Williamson, North and Ostrom (C-WON) allows the analyst to embrace complexity and elaborate a new calculus of institutional change based on the benefits and transaction costs for individuals and a 'winning coalition' needed for a rule change. This analytical perspective uses transactions as units of analysis, establishes models and typologies of transaction costs and institutional change, and enables the measurement and evaluation of institutional reforms in terms of adaptive efficiency.

Chapter 3 traced the evolution of river basin trajectories through the lens of path dependency and related transaction costs concepts: lock-in and intertemporal trade-offs. The development era in these rivers included three important institutional choices – irrigation supply organizations, water rights systems and interstate apportionment, and river basin governance arrangements – tailored to the values and preferences that prevailed at the time. These choices established the historical context for contemporary allocation reforms. This analysis illustrated the importance of vested interests and interdependencies, but also the potential for critical junctures or policy windows when this resistance can be overcome; when such windows open, the institutional design choices in cap-and-trade reforms aim to enhance flexibility and limit the incentives and tools available to block future change.

The three rivers experienced shifting fortunes in the ten years from 2004 to 2014, demonstrating that path dependency is not synonymous with stasis. In both water rights reform and river basin governance reforms, the political considerations and design choices interact to shape trajectories of reform. Reforms to river basin institutions have changed voting rules from interstate consensus to a unilateral central authority in the Murray–Darling in an effort to limit the veto powers of states and accelerate reforms, but may have proven counterproductive. By contrast, the steady, incremental and seemingly glacial reforms of the Colorado can yield non-linear progress, as illustrated in the binational breakthrough to restore water to the Colorado Delta with Minute 319, along with a flurry of side agreements since 2001.

Chapters 4 and 5 examined the emergence and maturation of water trading and basin governance in the Columbia and Murray–Darling, illustrating the importance of investments in institutional transitions and local governance capacity (Columbia) and polycentric governance arrangements (Murray–Darling). The local pockets of institutional reform and adaptive efficiency in the Columbia point to challenges for scaling up addressed in the Murray–Darling. Both rivers have essential ingredients for adaptive efficiency but would benefit from more effective coordination institutions (Columbia) and local capacity (Murray–Darling). Water trading and basin governance reforms are therefore part of Meinzen-Dick's 'institutional tripod' of states, markets and user associations, relying on institutional diversity in the water rights reforms and coordinating institutions to scale up (Meinzen-Dick, 2007).

These lessons illuminate a set of broader theoretical and practical implications for institutional analysis and development, namely the need to:

- embrace complexity and context;
- acknowledge that reform is never complete;
- design for diversity.

Embrace Complexity and Context: Large-Scale Collective Action and Mixed Property Systems

The decline of the Colorado River Delta, reduced outflows from the Murray Mouth and dewatered tributaries of the Columbia are symptoms of imbalances of supply and demand. River basin closure is a large-scale collective action dilemma (Marshall et al., 2013). This dilemma underscores water's complexity as an economic good with an array of interdependent public, common pool and private interests spanning from the household, farm, irrigation district and city to the state and federal governments, and the basin and wider regions as a whole. In this context, river basins are not the pre-given scale for resource management (Thiel, 2014); instead there are a group of linked action arenas with complex interdependencies across, within and beyond the upstream, downstream and basin-wide communities of interest. Addressing these dilemmas has required a recognition and embrace of the fundamental complexity, and hence politics, involved (Schlager and Blomquist, 2008). Free market environmentalism in its pure form – with its concept of atomistic and autonomous responses to price signals – has long been untenable, but we have lacked an alternative institutional form for embracing complexity. The linked concepts of mixed property systems, nestedness and polycentricity offer such an alternative. Water trading and basin planning can be viewed as part of a nested system of property rights and governance arrangements to reconcile local institutions for water allocation with the coordination institutions needed for larger-scale systemic trade-offs between interdependent values, users and jurisdictions. In this analysis, I have emphasized transaction costs analysis as a conceptual and empirical basis for understanding the co-evolution and performance of such nested systems.

Reform is Never Complete: A New Calculus for Transaction Costs and Institutional Change

In this context, we should not expect water allocation reform to be cheap. Nor should we expect all transaction costs to decline, because river basin closure heightens systemic risks and interdependencies. Embracing complexity therefore means embracing transaction costs. It means distinguishing different types of transaction costs and separating the necessary set-up costs incurred in the development of local water rights institutions and multilevel governance capacity from the costs imposed by the rent-seeking behavior of vested interest seeking to maintain the status quo. By merging the diverse traditions in transaction costs analysis, I have emphasized that not all transaction costs are created equal, and nor are all to be considered bad and in need of minimization. There are lock-in costs, transition costs and transaction costs, each with historical sources and different drivers and implications for institutional reform. The reduction of static transaction costs is often desirable and feasible with technological and institutional innovations, for example to create water rights registries, water banks, trading rules, and so on. Water trading requires that information is available to find buyers and sellers, assess the validity of property rights, assign a price, and monitor and enforce trades. Beyond these transactions, a wider set of linked transactions has proven essential to build and sustain institutional frameworks governing property rights and planning, and these are in turn shaped by historical forces that link past decisions with current and future trajectories of change. The pockets of success in the Columbia Basin illustrated the importance of upfront and ongoing investments in institutional transition costs to build adaptive efficiency; long before water trickled back to restore the rivers, the most successful cases used the set-up period to establish institutional frameworks to link multistakeholder water planning with water rights institutions to support market-based transactions (for example, water banking in the Deschutes). Those sub-basins that focused narrowly on the transaction hit the wall quickly after exhausting the limited opportunities available under the prevailing rules (Montana). Still, the larger-scale and basin-wide outcomes require coordinating institutions and integration mechanisms to address another set of transactions across jurisdictions (Colorado and Murray–Darling). Transition costs and impacts of lock-in costs are therefore empirical questions and less likely to decrease monotonically.

Solutions will be temporary, incomplete and often slow to develop (Blomquist, 2011; Briscoe, 2014); efforts to accelerate progress are risky and may prove maladaptive in the long run. The Murray–Darling Basin illustrated the dangers of hasty and unilateral action in the reforms enacted in 2007 when the throes of the Millennium Drought aligned with a political opening as the Howard government came to a close. Although the ensuing Water Act and basin planning generated major outcomes, the centralization of authority has triggered backlash and undermined the slow, but often effective, consensus-based interstate coordination that had unfolded over the previous century.

Designing for Diversity: Balancing Subsidiarity and Complementarity

In the era beyond panaceas, institutional diversity reigns (Ostrom, 2009). Property rights systems defy two-dimensional typologies based on exclusion and rivalry (Ostrom, 2010). Instead they are mixed assemblages of private claims, common pool resources and public goods nested across local, state, interstate and federal (and even international) authorities (Cole, 2002). Institutional and organizational arrangements for water rights reform and basin planning are hybrids with an array of roles and responsibilities assigned, based on an admixture of institutional fit and historical circumstances. The precise mix of water rights reform, community-based management and interstate coordination varies according to starting points and path dependencies.

In this context, principles of subsidiarity and complementarity offer guideposts for reform, namely to reserve or devolve tasks to the lowest level possible but no lower, while ensuring complementary coordinating institutions to make the trade-offs that span communities of interest (Marshall, 2005). In the Colorado, states' rights and the prominent role of the Bureau of Reclamation establish a dynamic tension between subsidiarity to respect allocation and planning at the state level with complementary basin-wide coordination. In the Columbia, states and sub-basins retain authority through state and field-based water rights administration and local watershed governance institutions, respectively; complementary basin-wide coordinating institutions are more limited despite the local presence in sub-basins with federal projects, and the non-binding roles of the Northwest Power and Conservation Council and Columbia Basin Water Transactions Program. The Murray–Darling has strived to achieve coherence amidst diversity. With constitutional provisions vesting allocation authority with the states, there has been a long history of consensus-based decision-making, with individual states wielding veto powers that limited the pace of progress and the implementation of intergovernmental agreements. Drought, environmental degradation and a new political will for reform prompted the establishment of both a new federal authority, the Murray–Darling Basin Authority, along with a counter-pressure for 'localism'. This has revealed the interdependency of informal institutions, such as local catchment organizations and ad hoc working groups for interstate coordination on the one hand, and the advent of new formal mechanisms for basin-wide planning and rulemaking, such as the Murray–Darling Basin Authority and Australian Consumer and Competition Commission, on the other hand. Designing for diversity requires mapping of institutional capacities and coordination gaps to ensure the appropriate assignment or reassignment of tasks, as well as the complementary institutions for addressing the lingering institutional collective action dilemmas among states.

POLICY IMPLICATIONS

Projected future development, demand and drought are putting these already stressed rivers under more pressure. Shortage on the Lower Colorado now looms as early as 2016, and the Millennium Drought (1997–2009) tested the resilience of the Murray–Darling and its water institutions. There is an increasing need to learn from the past and develop a comparative perspective about success and similarities, as well as failures and differences. Recent experiences with water reform in the three basins challenge conventional wisdom in the global water discourse in at least three areas.

First, the basin is not always the ideal unit for water management. Despite prescriptions for integrated water resources management (IWRM), progress has been elusive. IWRM requires greater focus on harnessing local capacity, and addressing the root causes of basin closure: vested interests of the irrigation development era. Basin-wide initiatives should provide incentives for management by stakeholders and users at all levels, particularly in the agricultural sector.

Second, there are no permanent solutions in an era of hard limits, systemic risks and tough trade-offs. Historic agreements have led to overallocation and path dependency. Recent droughts provide policy windows for creative solutions, but also opportunities for maladaptation. Systemic risks require the recognition of limits and the need for trade-offs; but there will be winners and losers, not always a win–win solution. Because solutions are temporary and consequences are lasting in a path dependent world, there is increasing need to learn from operating experience through a portfolio of decentralized (trading) and coordinated (for example, basin planning) tools, often adopted on an interim basis to gather learning experience.

Third, context matters, but every case is not (completely) unique. There has been a welcome push back against 'one-size-fits-all' approaches, but the pendulum has arguably swung too far. Although context matters, there is untapped potential to exchange lessons, and learn from success, failure, similarities and differences, particularly at the level of principles and institutional design. More rigor is needed, as is closer engagement and exchange by practitioners across the Colorado, Columbia and Murray–Darling. 'Diagnostic approaches' hold new promise in this regard by systematically unpacking complexity by decomposing large systems into their constituent subsystems, key variables, interactions and outcomes (Cox, 2011).

Learning from the Past: Pinpointing the Root Causes of Basin Closure

Recalling the opening chapter, the 'close parallels' between the people, culture and environment of the Colorado and Murray–Darling Basins have been the basis for a long history of exchange, dating at least to Alfred Deakin's Royal Commission study tour of the South West US in the 1880s. The Columbia is different in many respects, but its interior tributaries share several historical, cultural and environmental features with the

Colorado and Murray–Darling, each semi-arid tributary of the Columbia featuring a microcosm of the basin-wide characteristics in the other two basins. A major conclusion of the Royal Commission tour was to learn from both successes and failures of the budding US irrigation experiments. This led to strong state control over water development and water allocation in Victoria, Australia to avoid the apparent chaos and conflict in the late nineteenth century prior appropriation system (first in time, first in right) taking hold across the Western US.

The shared legacy did not stop there. Irrigation development remained closely linked through key individuals and the institutions they shaped. The Chaffey brothers from Canada met Deakin during his visit to California and travelled to Victoria to establish an ill-fated irrigation scheme near Mildura. George Chaffey then returned to California to help establish Imperial Valley, a linchpin in both California's and the Colorado River's water past and future. A less often noted fact is that Elwood Mead – the namesake for the Lower Colorado reservoir straddling the Arizona–California–Nevada border – became chair of the Victoria State Rivers and Water Supply Commission after his stint as Wyoming State Engineer and before his tenure as Commissioner of the US Bureau of Reclamation.

The imprint of these exchanges has been lasting. Contemporary efforts to adapt to climate change and competition across irrigation, urban, energy and environmental needs are fundamentally shaped by the strong vested interests and infrastructure commitments enacted by these early plans, laws and investments. In short, path dependency matters. The closure of both basins – the Colorado and Murray–Darling – was set in motion by the irrigation vision embodied by these early individuals and their exchanges. But the impacts took time to manifest, and by the time they did, it was arguably too late to avoid the political stalemates that hinder institutional reform today.

Overallocation, Systemic Risk and Adaptation: Redefining Limits in a Path Dependent World

Water rights and interstate apportionment agreements differ between the Murray–Darling, Colorado and Columbia, but sometimes the difference is overstated. Interstate apportionment agreements were needed to provide benefit- and cost-sharing arrangements before major infrastructures – dams and distribution – were constructed.

The River Murray Waters Agreement of 1914–15 represented an innovative approach, negotiated in the long shadow of the Federation Drought, which highlighted the need for risk sharing to cope with climatic variability. This led to proportional allocation systems (equal shares of available water) for the upper states of Victoria and New South Wales and fixed delivery to downstream South Australia. Not long after, the Colorado Basin states negotiated a Compact for water sharing; like the Murray, equitable use provided a guiding principle. However, the flooding of 1905, not only the droughts of the 1890s, was prominent in the minds of negotiators in a region exposed to climatic extremes. The Compact included fixed volumetric apportionments (rather than proportional), which placed disproportionate risk on the upstream states and fuelled legal disputes between the lower division states of Arizona and Colorado.

The Colorado River Compact overallocated the basin by promising more water than was available, and has become a poster child of 'stationarity' assumptions in water management, notorious for overestimating long-term average annual runoff and overlooking the potential for decadal droughts. But interstate water-sharing evolved in a similar direction as the Murray–Darling. The upper basin states agreed to a proportional allocation, while the lower basins secured fixed deliveries.

In both Australia and the Western US, states retain authority for water allocation. However, state water institutions pursued different systems of water rights, as shares (Australia) versus fixed priorities (Western US); however, even this difference is overstated in both legal and practical terms. The Australian systems of general and high-security entitlements provide a system of prioritization loosely akin to prior appropriation, and states jealously guard water from leaving their jurisdiction, much as those in the US.

In the US, the prior appropriation is often more rhetoric than operational reality. Federal water projects, which are particularly important in the Colorado River, guide deliveries for different classes of water which are treated as shares of available water within many irrigation districts, much like Australia's entitlement system. The common thread is that property rights are complex bundles of rights with key influences by irrigation institutions, state agencies and federal infrastructure projects.

The convergence of drought, demand and development exposed both regions to real limits. Droughts during the mid-twentieth century occurred when irrigation was dominant. By the 1980s, cities and the environment joined the party and heralded the era of limits and river basin closure. In the Murray–Darling this triggered moratoria in licenses at the state level as early as the 1960s (South Australia) and became basin-wide after the 1994 Council of Australian Governments (COAG) reforms and the ensuing interim cap.

In the Colorado River, the tremendous buffering capacity of reservoir storage (a 4:1 ratio of storage to annual runoff) postponed this reckoning until 1999 when long-range demand and supply intersected. An unprecedented and ongoing sequence of dry years led to demand exceeding long-range supply by 2002 and has depleted reservoirs.

In the Columbia River, the decline of salmon fisheries is attributed to many factors. Yet irrigation diversions have dewatered upstream tributaries where salmon reproduce, exposing physical limits as early as Oregon's 1955 Minimum Perennial Streamflow Act. A combination of court decrees and administrative rules, including those establishing minimum flows, have been used to establish an uneven patchwork of limits.

The responses to the era of limits have therefore diverged, in large part owing to differences in both the perception and the reality of scarcity. The Murray–Darling has adopted an interim cap, followed by sustainable diversion limits, backed by associated audits, plans and commitments to return the basin to sustainable levels of extraction in the 2004 National Water Initiative and the 2007 Water Act. Water markets have provided a key vehicle for reallocation, and federal and state programs have recovered water for the environment totalling more than 1500 gigalitres (GL). Although water for cities has become an important demand on the river for Adelaide, Canberra and Melbourne, the Murray–Darling adjustment to a future of less water has played out as a clash of regional irrigation communities and the environmental needs of wetlands across competition state jurisdictions.

In the Colorado River, the era of limits has prompted a wave of reforms, starting ironically with the management of surpluses in 2001–02 to address California's historic use above its legal allocation. Shortage planning and coordinated reservoir management was the focus from 2004 to 2007. The 2010–12 Basin Study involved a federal–state cost share to undertake system–wide basin planning and identify and respond to projected average annual supply–demand imbalances totalling up to 3.2 million acre feet annually by 2060. The results of the study included a raft of management options ranging from supply-side measures (importation, cloud seeding, desalination) to demand side measures (conservation, transfers, and so on). The process revealed a fundamentally different approach to the Murray–Darling: one based on extrapolating demand rather than acknowledging a limited long-term supply. However, even these fundamental differences mask important similarities, as evidenced by the predilection for irrigation efficiency schemes as a means for achieving water recovery objectives.

The common challenge for both basins is that river basin closure reveals systemic risks and increasing interdependencies between sectors, jurisdictions and communities. Solutions – whether for demand imbalances or environmental recovery – are temporary, but the consequences are lasting.

NAVIGATING FUTURE TRADE-OFFS: DIAGNOSTIC AND COMPARATIVE RESEARCH

Improving water security during droughts and addressing neglected environmental and indigenous water needs will not be cheap – politically or economically. Further, there will be no uniform, one-size-fits-all solution. At the level of policy implications, some important parallels and implications are evident. To recap:

1. The importance of recognizing limits and dealing with climate extremes. Each basin is chronically overallocated, but only the Murray–Darling has recognized the need to adopt sustainable diversion limits. Decadal droughts create policy windows for lasting change, but also opportunities for maladaptation.
2. Solutions are temporary; consequences are lasting. As such, systemic crisis requires systemic planning and adaptation with a polycentric mixture of solutions.
3. Key governance roles and leadership needs exist at every level, from local to federal and international. A program of research is needed to understand and analyze efforts to design and develop institutional diversity.
4. Comparative research and practice provide lessons from success and similarities, and from failure and differences. Comparative methodology needs to be rigorous and interdisciplinary and to bridge disciplinary divides and the gaps between research and practice divide.

On the surface, one could argue that there are too many differences to learn from comparison across the three basins. But this takes a narrow view of comparative research. Wescoat (2009) reminds us that:

> The 21st century will almost certainly be an era of increased global circulation of water issues, inquiry, expertise, and action . . . in light of the critical water problems faced in every region of the world, the next twenty years will require a major shift from largely implicit comparisons to rigorous comparative analyses.

The bilateral flow of researchers, practitioners and ideas across the basins is truly striking. The future agenda for research and practice must embrace this process to deepen comparisons and improve methods for doing so, building on diagnostic approaches that systematically define the relevant questions and variables to enable comparative and longitudinal assessment. After over a century of exchange, we have discovered that we have only scratched the surface. For three basins positioned at the leading edge of global change, it is time to learn more from the past and provide a sound platform for mutual learning about the options for navigating future trade-offs in a transaction costs world. Path dependency will require care in transferring more specific lessons in other historical, biophysical and development conditions.

Reflecting on the two institutional panaceas – water markets and river basin governance – examined in this book, it is fitting to close with the observations of Garrido (2011) and Meinzen-Dick (2007). There are no panaceas, and the approaches that work in one setting may not work in another, unless adapted carefully to local circumstances. To paraphrase Garrido, the opposite is also true, namely that institutional failure in one river does not preclude success in another. A focus on adaptive efficiency, and the path dependent relationship of transaction costs and institutional change, suggests that successful efforts will depend on institutional diversity and sustained, multilevel investments in collective action.

REFERENCES

BLACKBOURNE, D. 2006. *The Conquest of Nature: Water, Landscape and the Making of Modern Germany*. London: Norton.

BLOMQUIST, W. 2011. A political analysis of property rights. In: COLE, D.H. and OSTROM, E. (eds), *Property in Land and Other Resources*. Cambridge, MA: Lincoln Institute of Land Policy.

BRISCOE, J. 2014. The Harvard Water Federalism Project – process and substance. *Water Policy*, 16, 1–10.

CAREY, J.M. and SUNDING, D.L. 2001. Emerging markets in water: a comparative institutional analysis of the Central Valley and Colorado–Big Thompson Projects. *Natural Resources Journal*, 41, 283–328.

COLE, D.H. 2002. *Pollution and Property: Comparing Ownership Institutions for Environmental Protection*. Cambridge: Cambridge University Press.

COOK, B.I., AULT, T.R. and SMERDON, J.E. 2015. Unprecedented 21st century drought risk in the American Southwest and Central Plains. *Sci. Adv.*, 1, no. 1, e1400082, doi:10.1126/sciadv.1400082.

COX, M. 2011. Advancing the diagnostic analysis of environmental problems. *International Journal of the Commons*, 5, 346–63.

CULP, P., GLENNON, R. and LIBECAP, G. 2014. Shopping for water: how the market can mitigate water shortages in the American West, 1–8.

GARRICK, D., WHITTEN, S. and COGGAN, A. 2013. Understanding the evolution and performance of water markets and allocation policy: a transaction costs analysis framework. *Ecological Economics*, 88, 195–205.

GARRIDO, S. 2011. Governing scarcity: Water markets, equity and efficiency in pre-1950s eastern Spain. *International Journal of the Commons*, 5(2), 513–34.

GLENNON, R. and LIBECAP, G. 2014. Opinion: The West needs a water market to fight drought. *Wall Street Journal*. Published October 23, 2014.

Available at: http://www.wsj.com/articles/robert-glennon-and-gary-libecap-the-west-needs-a-water-market-to-fight-drought-1414106588.

GRAFTON, R.Q. and HORNE, J. 2014. Water markets in the Murray–Darling Basin. *Agricultural Water Management*, 145, 61–71.

KAUFMAN, G.G. and SCOTT, K.E. 2003. What is systemic risk, and do bank regulators retard or contribute to it? *Independent Review*, 7(3), 371–91.

KENNEY, D.J. 2009. The Colorado river: what prospect for 'a river no more'. In: MOLLE, F. and WESTER, P. (eds), River Basin Trajectories: Societies, Environments and Development. Colombo, Sri Lanka: IWMI.

MALLOCH, S. 2005. Liquid Assets: Protecting and Restoring the West's Rivers and Wetlands through Environmental Water Transactions . Arlington, VA: Trout Unlimited, Inc.

MARSHALL, G. 2005. *Economics for Collaborative Environmental Management: Regenerating the Commons*. London: Earthscan.

MARSHALL, G., CONNELL, D. and TAYLOR, B. 2013. Australia's Murray–Darling Basin: a century of polycentric experiments in cross-border integration of water resources management. *International Journal of Water Governance*, 1, 197–218.

MEINZEN-DICK, R. 2007. Beyond panaceas in water institutions. *Proceedings of the National Academy of Sciences*, 104, 15200–205.

NATIONAL WATER COMMISSION (NWC). 2011. *Strengthening Australia's Water Markets*. Canberra, ACT: NWC.

OSTROM, E. 2009. *Understanding Institutional Diversity*. Washington, DC: Princeton University Press.

OSTROM, E. 2010. Beyond markets and states: polycentric governance of complex economic systems. *American Economic Review*, 100, 641–72.

ROBINSON, C.J., BARK, R.H., GARRICK, D. and POLLINO, C.A. 2014. Sustaining local values through river basin governance: community-based initiatives in Australia's Murray–Darling basin. *Journal of Environmental Planning and Management*, 1–16.

SCHLAGER, E. and BLOMQUIST, W.A. 2008. *Embracing Watershed Politics*. Boulder, CO: University Press of Colorado.

THIEL, A. 2014. Rescaling of resource governance as institutional change: explaining the transformation of water governance in southern Spain. *Environmental Policy and Governance*, 24, 289–306.

US BUREAU OF RECLAMATION. 2012. *Colorado River Basin Water Supply and Demand Study*. Washington, DC: US Department of the Interior.

Index